WITHDRAWN

OTHER TITLES OF INTEREST FROM ST. LUCIE PRESS

Principles of Sustainable Development

An Introduction to Environmental Economics

Multiple Objective Decision Making for Land, Water and Environment

Environmental Management Tools on the Internet: Accessing the World of Environmental Information

Privatization of Information and Agricultural Industrialization

Ecological Integrity and the Management of Ecosystems

Ecology and Management of Tidal Marshes: A Model from the Gulf of Mexico

Everglades: The Ecosystem and Its Restoration

The Everglades Handbook: Understanding the Ecosystem

The Everglades: A Threatened Wilderness (Video)

Ecology and the Biosphere: Principles and Problems

Resolving Environmental Conflict: Towards Sustainable Community Development

Economic Theory for Environmentalists

Development, Environment, and Global Dysfunction: Toward Sustainable Recovery

From the Forest to the Sea: The Ecology of Wood in Streams, Rivers, Estuaries, and Oceans

Environmental Fate and Effects of Pulp and Paper Mill Effluents

Environmental Effects of Mining

Sustainable Community Development: Principles and Concepts

Environmental Policy and NEPA: Past, Present, and Future

Wildlife Issues in a Changing World

For more information about these titles call, fax or write:

St. Lucie Press
2000 Corporate Blvd., N.W.
Boca Raton, FL 33431-9868

TEL (561) 994-0555 • (800) 272-7737
FAX (800) 374-3401
E-MAIL information@slpress.com
WEB SITE http://www.slpress.com

S$^t_{L}$

Environmental Sustainability
Practical Global Implications

Edited by
Fraser Smith

CARL A. RUDISILL LIBRARY
LENOIR-RHYNE COLLEGE

S^{t}L

St. Lucie Press
Boca Raton, Florida

HC
59.72
.E5
E 627
1997
Oct.1997

Copyright ©1997 by CRC Press LLC
St. Lucie Press is an imprint of CRC Press

ISBN 1-57444-077-2

Printed in the United States of America   10   9   8   7   6   5   4   3   2   1
Printed on acid-free paper

   This book contains information obtained from authentic and highly regarded sources. Reprinted material is quoted with permission, and sources are indicated. A wide variety of references are listed. Reasonable efforts have been made to publish reliable data and information, but the author and the publisher cannot assume responsibility for the validity of all materials or for the consequences of their use.

   Neither this book nor any part may be reproduced or transmitted in any form or by any means, electronic or mechanical, including photocopying, microfilming, and recording, or by any information storage or retrieval system, without prior permission in writing from the publisher.

   All rights reserved. Authorization to photocopy items for internal or personal use, or the personal or internal use of specific clients, may be granted by CRC Press LLC, provided that $.50 per page photocopied is paid directly to Copyright Clearance Center, 27 Congress Street, Salem, MA 01970 USA. The fee code for users of the Transactional Reporting Service is ISBN 1-57444-077-2/97/$0.00/+$.50. The fee is subject to change without notice. For organizations that have been granted a photocopy license by the CCC, a separate system of payment has been arranged.

   The consent of CRC Press does not extend to copying for general distribution, for promotion, for creating new works, or for resale. Specific permission must be obtained from CRC for such copying.

   Direct all inquiries to CRC Press LLC, 2000 Corporate Blvd., N.W., Boca Raton, Florida 33431.

For Laura, to whom it's all obvious

Contents

Foreword

In the hue and cry over the need to become more sustainable in our brief tenancy on this unique and miraculous planet, it is all too easy to ignore the voices of those who can teach a lot about what it is like to live sustainably. Throughout the globe, there are heartening attempts to create meaningful arrangements in environmental rights and communal obligations that can lead to a more enduring habitation. Why do we not listen to such voices? Mostly because we do not appear to want to learn, but also because there is no forum for their articulation.

Fraser Smith and his colleagues address this second point. What follows is an unusual collection of essays, primarily rooted in Third World experience, where the principles of living ecologically are converted into economic and social experience. It should come as no surprise, but nevertheless it is still worthy of note, that societies that exist to care for their surroundings and for each other appear more healthy, happy, and forgiving. The paradox is that the greater the accumulation of wealth, despite its hugely damaging repercussions, the greater the avarice and the less happy or generous are the accumulators. This is a frightening conclusion. Sustainable development means sharing and caring for a humanity that has to tend for the planet as well as for itself. Yet democratic institutions are built up on patterns of power that only gain their influence because of exploitation of nature and peoples. So how do we shift toward sustainability through a democracy that is dependent on unsustainable practices?

The answer is by reading and responding to the heartening chapters that follow. These are the voice of pain, suffering, yet also the voice of the joyous innocence of the pre-exploited Amerindians to the brave initiatives by non-governmental organizations to retain or return property rights into common ownership and reciprocal obligation. Much of these imaginative schemes are being tried out against the grain of development pressures, international aid, and national government corruption and patronage practices which distort all prices to featherbed the already wealthy.

This is a book ostensibly about ecological economics, an emerging interdisciplinary science that seeks to unite the home of humankind to the home of nature. Natural systems provide humanity with a vast array of environmental services that we are beginning to understand, codify, and value. The important task of the ecological economist is to ensure that these vital functions are fully recognized before, not after, development has taken place, so that the well-being of natural systems and the quasi-sustainable human communities that maintain them are a source of both admiration and respect, as well as a vital laboratory for the future.

Yet the "pain" of damage lies rooted in so many hostile political, economic, and social arrangements the world over. We cannot begin to put the global house in order before we turn to these hugely damaging arrangements and, with the persistent voice of eco-sanity, begin to dismantle them. This book helps us all to see how it can be done and that the task is so desperately important for us all.

Timothy O'Riordan
School of Environmental Sciences
University of East Anglia
Norwich, U.K.

Preface

However much you study you cannot know without action.

Thirteenth-century Sufi author Saadi of Shiraz

While sustainable development has become an increasingly important issue in academia, government, and business, the vast majority of discussions on the subject that reach a wide audience emanate from the rich, industrialized nations. These discussions have mostly ignored the important contribution from developing countries, whose authors often have distinct and revealing ways of approaching the subject. This book aims to redress this imbalance in the literature by presenting the state of the art on sustainable development from outside the countries of the Organization for Economic Co-operation and Development (OECD). Most of the chapters in the book discuss practical approaches that can be, or are already being, taken to implement environmentally sustainable economies. This emphasis is the rationale for the book, based on the old adage that actions speak louder than words, and therefore broadcasting the actions rather than the words will add the greatest value. However, as this book reveals, there are important philosophical differences between the "Northern" and "Southern" notions of sustainable development, and therefore an equally important task here is to set the basis for a more complete vision combining the two perspectives.

The idea for the book originally came out of the second biennial meeting of the International Society for Ecological Economics in Stockholm in 1992. That meeting itself lacked significant input from developing countries. Many participants, myself included, felt that improving the ISEE's representation from outside the OECD would be a valuable exercise. The subsequent meeting in Costa Rica in 1994 generated a profusion of ideas among an extremely diverse mixture of delegates from all over the developing world, and the chapters herein are largely derived from papers presented at that meeting. The contributors are drawn from universities, research institutes, governmental organizations, and non-governmental organizations. They are mostly nationals of developing countries plus a few nationals of OECD countries who have spent substantial amounts of time working in the developing world.

The perspectives presented here are drawn from Central and South America, sub-Saharan Africa, and South and Southeast Asia. They cover an extremely broad range of topics, including the philosophy of sustainable development, institutions, ethics, belief systems, "indigenous" cultures, resource use, energy use, economic modeling, econometric analysis, international trade, financial aid, forestry, wildlife, land rights, fisheries, and more. Whatever the interest of the reader on sustainable development, this book is likely to contain something useful.

The breadth of the book was an explicit aim from the beginning, for two reasons. First, none of the chapters that follow is intended as an exhaustive in-depth analyses of its topic; nonetheless, analytical rigor is still the first priority. The chapters present familiar topics in a new light, suggesting ways to improve current theory and practice and proposing avenues for further work. The second reason for a broad scope is that the book is intended for a wide audience. Catering to a diverse readership will, it is hoped, give the "Southern" perspective the prominent voice it deserves. In preparing this book, and at the conference that spawned it, I was surprised by how strongly colleagues from developing countries felt about increasing the profile of the Southern view in the international arena. To this end, the book should be readable by academics and interested lay parties alike; both can probe further into particular areas as their interest takes them. Technicalities—equations, acronyms, and jargon—are confined as far as possible to footnotes.

The value of this volume may be measured not only in terms of the ideas that it contains but also as a reference guide. Students in natural and social sciences may find material for writing papers, development professionals may find ideas for improving their operations, entrepreneurs may find opportunities for ecologically sustainable businesses, and academics may make contact with colleagues they would not normally have found. It is hoped that, as a source of contacts especially, the book will not only bring workers from rich and poor countries into collaboration but will also bring workers together from different parts of the developing world, which is often a difficult task.

The book is divided into two major sections which address the dual aims of the book described above. The first discusses philosophical and conceptual aspects of sustainable development from the point of view of people in developing countries. One of the more prominent themes in this section is the need to change attitudes in the wider world to be more closely aligned with those people who live in relatively sustainable harmony with their natural surroundings. Clóvis Cavalcanti draws contrasts between the attitudes and lifestyles in the United

States and those of the Amerindians in Brazil and argues that while the Amerindian lifestyle is "not a panacea," it serves as an example of a set of guiding principles to bring the rich countries closer to ecological sustainability. At the same time, it might bring greater fulfillment to the people of the rich countries, whose social dislocation is clearly evident. The authors of these preliminary chapters universally agree that global environmental sustainability is possible only if people in rich countries change the way they live, and the authors offer some guides on how to do this.

The second section contains detailed case studies from around the developing world that are used to support new arguments about sustainable development or test existing ones. Many of these case studies highlight the challenges and successes of specific efforts to make economic activities more ecologically sustainable. Many of these chapters tackle head-on the linkages among social and ecological processes operating at a variety of levels. Local ecological conditions and economic practices in developing and industrialized regions influence each other via a variety of large-scale natural and economic forces. In many cases, social and ecological problems arise from distributional inequities between rich and poor and in the geographical discounting of environmental and social degradation. The authors present a diverse array of approaches—institutional, political, and market based—to improve the way these linkages operate.

At the Costa Rica meeting from which this book is drawn, there was much talk of "visioning" how the world should be in fifty, a hundred, or a thousand years. Not surprisingly, people's visions converged on a world in which humanity lives in peace and comfort within natural limits. This volume presents in its first part a collective vision that attempts to distill how the world's poor would like things to turn out and in its second part a vision on how to get there. These visions are perhaps most valuable for environmental sustainability because the usual emphasis on material well-being is counterbalanced by spiritual and social well-being, which cannot be valued monetarily.

The appeal of the book is, it is hoped, as much in the details as in the broader messages. The wealth of field data collected in the following pages may stimulate the reader to concoct new avenues of investigation into sustainable development or to design novel practical applications. While the overriding message is that the South has much of importance to say about sustainable development which could or should be acted upon, the underlying message is that developing countries also harbor much of the necessary knowledge and expertise, until now barely tapped. Readers in rich and poor countries alike will, I hope, be stimulated enough by the results from others' backyards to try it in their own. At any rate, whether as a source of reference or as a philosophical statement, I hope the reader will find the book as much of an eye-opener as I have in the course of its preparation.

Finally, grateful acknowledgments go to Ed DeBellevue for his valuable help and advice in the early stages of this project and to Sandra Koskoff for facilitating the partnership with St. Lucie Press. Special thanks go also to all the contributors, whose commitment to the project and diligent communication across great distances went far beyond my expectations.

Fraser Smith
San Francisco

The Authors

Kathleen L. Abdalla is First Economic Affairs Officer at the United Nations Economic and Social Commission for Western Asia (Amman, Jordan), P.O. Box 5749, New York, New York, 10163–5749, USA. E-mail: kathleen@ nets.com.jo.

Neil Adger is a Senior Research Fellow at the Centre for Social and Economic Research on the Global Environment, University of East Anglia, Norwich NR4 7TJ, and University College, London, UK. E-mail: n.adger@uea.ac.uk.

Juan Antonio Aguirre Gonzalez is a Graduate Teaching and Research Professor in Agriculture and Environmental Economics and Head of Graduate Studies at the Centro Agronómico Tropical para Investigaciónes y Enseñanza (CATIE), Apartado 52-7170, Turrialba, Costa Rica. E-mail: jaguirre@catie.ac.cr.

Mahfuzuddin Ahmed is a Social Scientist at the International Center for Living Aquatic Resources Management (ICLARM), MC P.O. Box 2631, 0718 Makati City, Philippines. Fax: 63 2 816 3183.

Caroline Ashley is a Resource Economist at the Directorate of Environmental Affairs, Ministry of Environment and Tourism, P.O. Box 13306, Windhoek, Namibia. E-mail: ca@dea1.dea.met.gov.na.

Jonathan Barnes is a Resource Economist with the World Wildlife Fund (U.S.) LIFE Programme, Directorate of Environmental Affairs, Ministry of Environment and Tourism, P.O. Box 13306, Windhoek, Namibia. E-mail: jb@dea1.dea.met.gov.na.

Roxana M. Barrantes is a National Consultant for the Organismo Supervisor de la Inversion Privada en Telecomunicaciones, Centro Comercial Camino Real, Torre El Pilar, Lima 27, Perú, and an Assistant Professor at the Department of Economics, Pontificia Universidad Católica del Perú. E-mail: rbarrant@ pnud12.pnudreg.org.pe.

Ana Doris Capistrano is a Program Officer at the Ford Foundation (India Office), 55 Lodi Estate, New Delhi 110003. E-mail: d.capistrano@fordfound.org.

Clóvis Cavalcanti is a Senior Researcher at the Fundação Joaquim Nabuco, INPSO, R. Dois Irmaos, 52071-440 Recife, PE, Brazil. E-mail: clovati@fundaj.gov.br.

Kanchan Chopra is a Professor at the Institute of Economic Growth, University Enclave, Delhi 110007, India. E-mail: kc@ieg.ernet.in.

Anil K. Gupta is a Professor at the Centre for Management in Agriculture, Indian Institute of Management, Ahmedabad 380015, India. He is also the Coordinator of SRISTI (Society for Research and Initiatives for Sustainable Technologies and Institutions) and Editor of *Honey Bee,* a magazine on sustainable development. E-mail: anilg@iimahd.ernet.in.

Kiew Bong Heang is a Lecturer at the Department of Zoology, Faculty of Science, University of Malaya, 59100 Kuala Lumpur, Malaysia. E-mail: kiew@zoology.um.edu.my.

Mokammel Hossain is Project Director of the Community-Based Fisheries Management Project, Department of Fisheries, Government of Bangladesh Ministry of Fisheries and Livestock, Park Avenue, Ramna, Dhaka, Bangladesh.

Aseem Prakash is an Assistant Professor at the Department of Strategic Management and Public Policy, School of Business and Public Management, The George Washington University, Washington, D.C., USA.

Fraser Smith is an Infoware Manager at Datafusion, Inc., 3220 Sacramento Street, San Francisco, CA 94115, USA. E-mail: fraser@datafusion.net.

Victor M. Toledo is a Researcher at the Centro de Ecología, Universidad Nacional Autónoma de México, Apartado 41-H Santa Maria Guido, Morelia, Michoacán 58090, Mexico.

A Synthetic Framework and a Heuristic for Integrating Multiple Perspectives on Sustainability

1

*Fraser Smith**
Datafusion, Inc., San Francisco, California

Abstract

Because the challenge of sustainability is heterogeneous in time and space, it will require heterogeneous solutions. This introductory chapter draws together the findings of the other chapters in the present volume to draft a synthetic framework to guide this heterogeneous challenge. The inspiration for this approach comes from the synthetic theory of biological evolution, which emerged from two divergent evolutionary disciplines in the 1930s and 1940s.

The framework integrates the heterogeneous "Southern" perspectives presented in subsequent chapters with the predominating "Northern" perspective characterized here. The Southern views contain a number of common themes, elaborated in this chapter, which differ substantially from the Northern perspective. The present chapter makes a critical analysis of both sets of views and integrates them for a more complete vision, establishing broad criteria for sustainability and enumerating a suite of metrics. Under the framework, different criteria and different metrics will apply in different circumstances. As an illustrative example, the framework is applied to the problem of harvesting fish for ecological stability and economic return.

---

* The views expressed in this chapter are those of the author and not necessarily those of Datafusion, Inc.

1-57444-077-2/97/$0.00+$.50
© 1997 by CRC Press LLC

1

One of the recurring themes in this book is the emphasis on economic growth as a *prerequisite* of sustainability in developing countries. Using a simple heuristic to model economic activity in relation to resource throughput, population, and technology, it is shown that sustainability would allow economic growth as long as certain conditions were satisfied concerning technology; in fact, certain technologies may be beneficial for sustainability. It is intended that the use of heuristics, such as the one presented here, may, within a synthetic framework for sustainability, help to identify the important drivers for formulating integrated, heterogeneous solutions to the problem.

> Dominion of the world from end to end
> Is worth less than a drip of blood upon the earth.
>
> *Saadi of Shiraz*

Introduction

In the last few years, much has been written about the concept of "sustainable development," to the point where whole books are devoted to defining it (e.g., van den Bergh and van der Straaten, 1994; Reid, 1995). The term is, in fact, so vague that it has been used not only by advocates of precaution to refer to the environmental sustainability of economic activity but also by advocates of growth to refer to the sustainability of economic expansion—two concepts that appear at first glance to be diametrically opposed. In 1987, the World Commission on Environment and Development (WCED) attempted to provide a definition of environmentally sustainable development which has, almost ten years hence, passed into common parlance: "sustainable development is development that meets the needs of the current generation without compromising the needs of future generations" (WCED, 1987). Unfortunately, this definition is conceptually flawed because it is impossible to know what the preferences of future generations will be. Logically, then, precaution would dictate the preservation of the natural environment in its unaltered state, and we thus arrive at the so-called "strong" definition of sustainability, which is economic development that does not compromise environmental integrity.* This form of sustainability is probably the most appropriate long-term policy goal (see Smith, 1996a).

---

* The WCED definition is sometimes referred to as the "weak" definition of sustainable development. This is distinct from the "weak sustainability" criterion of neoclassical–classical economics that permits natural and financial capital to be substitutable, on the assumption that the price of the natural capital reflects its true environmental value (which is rarely the case). The weak definition of sustainable development is more stringent than weak sustainability but may allow it if the price assumption is satisfied. See Gowdy and O'Hara (1995).

In fact, the strong definition of sustainable development can be justified in a second way, more related to current economic conditions. Contrary to prevailing assumptions, the strong definition need not exclude economic growth, as long as that growth is directed toward conserving environmental integrity. Put another way, the "expansionist" interpretation of sustainable development is essentially containable within the strong definition of sustainability, subject to constraints on exactly what is expanded. This interpretation of sustainability has plenty of empirical evidence to support it, but until now very little such evidence has been presented in one place. Much of it comes from developing countries, where economic growth is less universally regarded as the enemy of environmental integrity than in industrialized countries. The present volume describes practical applications toward sustainability emanating from developing countries, informing the concept in ways that are unfamiliar in the industrialized world. It also presents the philosophical underpinnings of sustainability from a developing country point of view and highlights the differences between this view and the "Northern"* one. Because substantial differences exist between these two points of view, the aim of the present chapter is to use the findings in the rest of the book and elsewhere to begin a synthesis of "Northern" and "Southern" perspectives into a more all-encompassing conception of sustainable development. Of course, if sustainability were achieved, we would probably not be talking about "North" and "South" anyway, but the fact is that pronounced distributional inequities exist between the higher income and lower income countries, as well as significant differences in resource intensity, attitudes toward the environment, and so on. These differences are what motivate the present discussion, but it is hoped that the exercise of an integrated conception of sustainability will eventually eliminate such labels as "North" and "South." The motivation for the present synthesis is that it might make the operation of international development projects more successful at improving the lives of the people they are supposed to help and that it might address fundamentally why the North has been so slow to become environmentally efficient.

The inspiration for this synthesis comes from evolutionary theory. In the early part of the 20th century, biologists were deeply split over the importance of natural selection as a driving force for evolution. The "naturalists" believed that natural selection was the only important evolutionary force and argued that it did not require a genetic basis. The "geneticists," by contrast, believed that the only significant force for evolution was genetic mutation: if a mutation had a large enough effect, it would bypass the incremental changes hypothesized by Darwin. In the 1930s and 1940s, a new view emerged that reformulated the

---

* "Northern" is defined here as countries belonging to the Organization for Economic Co-operation and Development (OECD) and "Southern" as non-OECD countries.

theory of natural selection with a strictly genetic basis. Certain other types of evolution were also hypothesized and later supported by empirical evidence (see Mayr, 1982). Known as the "synthetic" theory of evolution, this formulation has persisted largely intact to the present. That the synthetic theory took 20 years to reach maturity and broad scientific acceptance should make it quite clear that the present synthesis is strictly preliminary. The intention here is to focus attention on the important questions in sustainable development in different places, at different times, and under a variety of circumstances—just as the synthetic theory of evolution provides a broad framework for testing and interpreting biological phenomena. The real value of a synthetic conception of sustainability is its ability to shed light on how to make the best use of all available opportunities. Making sustainability operational is really a matter of predicting and measuring it, far more than just defining it (Costanza and Patten, 1995).

In addition to a synthetic framework, this chapter also presents a simple heuristic to explore from basic principles the conditions required for environmentally sustainable economic development. The main insight of the heuristic is in accord with the synthetic framework: sustainable development may encompass a variety of processes, in different places at different times, including some not usually associated with the conservation of environmental integrity. In particular, economic growth may in fact be not only compatible with sustainability but actually beneficial for it, and the main driving force for sustainability relates to the universe of human technologies.

It is hoped that these two complementary results will provide a powerful impetus for identifying the range of opportunities and constraints for sustainable development on a practical basis as well as a conceptual one. It is also hoped that the *diversity* of theoretical and practical approaches contained herein will make clear the *necessity* of heterogeneous progress toward sustainability. As Kaufmann and Cleveland (1995) correctly point out, "ecological economists need to graduate to a less aggregated, more interdisciplinary and more sophisticated notion of sustainability."

The present chapter is organized into five subsequent sections. First, the "Northern" perspective on sustainable development is briefly characterized and its main conceptual and practical deficiencies presented. A few brief points are made about why the Northern perspective has dominated efforts to institute sustainability and why it may be incomplete. Second, the main components of a range of contrasting Southern perspectives are presented, drawing on the work in subsequent chapters. These perspectives are critiqued from two Northern standpoints: one expansionist, one precautionist. These two sections then lead into a third, which synthesizes the foregoing perspectives into a conceptual framework and a set of practical prescriptions that address the problem of sustainable development. Fourth, the heuristic for identifying allowable conditions for sustainability is presented, and its insights are placed within the synthetic formulation. Finally, a set of generalized "signposts" is developed for decision making in a sustainable economy.

Sustainable Development as Envisioned in the North

Martinez-Alier, writing ten years ago, expressed surprise that there are "almost no ecological social movements with roots in the Third World." In almost the same breath, he expresses puzzlement that "left-wing ecologism has grown...not so much in the Third World as among part of the youth of some of the most over-developed countries" (Martinez-Alier, 1987, pp. 237, 238). The ecological critique of neoclassical economics was already under way in the industrialized countries almost 30 years ago (Boulding, 1966; Daly, 1968; Georgescu-Roegen, 1971), yet only more recently, and especially in the last 10 years, has a strong "ecopopulism" (Martinez-Alier's term) emerged outside the OECD countries (e.g., Cavalcanti, Chapter 2).

The notion of environmentally sustainable development was promoted in the 1970s most prominently by Herman Daly (1972), who argued that economies should not grow but exist in a dynamically steady state within environmental limits. This is essentially the strong definition of sustainability given previously. To move the debate from the academic to the political arena has, however, required a more politically expedient interpretation of the goal, which is encapsulated in the WCED definition. Many ecological economists maintain that this intergenerational form of sustainability should be treated as really no more than a stepping-stone toward the stricter biophysical form (e.g., Smith, 1996a). However, the fact remains that "sustainable development" as a concept is a product of the North, and this fact prompts two important questions:

1. Why did a similar concept not appear in the South?
2. Is the notion of sustainable development applicable to the South?

Before answering these questions, it is necessary to identify what constitutes sustainable development as seen through Northern eyes. Many of the requirements of sustainability in the Northern vision are replicated in the Southern perspective, as we shall see below, but the emphasis is different. Not all people in the North who believe in sustainable development would necessarily subscribe to the all of following criteria, but by definition they should subscribe to at least one:

1. The intergenerational requirement should be satisfied and, in addition, the more stringent requirement of not breaching biophysical limits should be achieved as soon as possible, regionally, globally, and continuously.

2. The economy should not grow in size, or at most should grow only by a very small amount, over the long term.
3. Discount rates should be abolished in the economy so that a long-term perspective is fostered.
4. Distributional equity should be encouraged, as should a conservationist ethic.
5. Industries and products should be environmentally non-damaging or beneficial, and individuals should choose their professions likewise.
6. Institutional and political changes should be enacted that foster changes in individual attitudes and behaviors toward environmental sustainability.
7. Economic instruments (taxes, quotas, etc.) should be used to regulate economic activity toward the sustainable goal.

Of these criteria, the one that has most pervaded the popular consciousness relates to the environmental soundness of products and the industrial processes that generate them. The label "environmentally friendly" is often financially lucrative for Northern manufacturers, even if a product does not live up to the billing. Conservationist concerns are now widely voiced among the populations of the rich countries, and in some (e.g., the Netherlands), sustainability is beginning to appear in lawmaking. However, most people do not think very much about intergenerational or broader biophysical criteria for sustainability as they relate to everyday life—even though these criteria alone could guide all the others—and attempts by administrations to foster distributional equity have had only limited success, at great fiscal cost. There is no sign of an imminent abolition of discount rates, nor have lawmakers had the courage to alter tax structures so that only environmental "bads" are taxed, and not income or other goods. As for the criterion on economic non-growth, it will be discussed later.

In answer to Martinez-Alier's question about why a concept of sustainable development did not appear in the South, the answer is probably that this is due to differences in living standards between South and North. Martinez-Alier himself notes in relation to his own "German" political ecologism that "there is some inconsistency about caring for the conservation of world resources while enjoying the average standard of living of prosperous Frankfurt, Amsterdam, or Berlin" (Martinez-Alier, 1987, p. 237). From a simple biological standpoint, once the immediate, internal needs of the individual are assured (i.e., at the most basic level, food and shelter), attention turns to external, more long-term issues (at the basic level, reproduction). In the countries with high average standards of living, it is hardly surprising that some people, after a generation of relative peace since 1945, would become motivated about the poor state of the natural environment and the economic structures that produced it. In the poor countries, by contrast, many people during this period had little time to worry about the global environment in the face of the struggle to subsist. This explanation, above all others, accounts for most of the differences between the Northern and Southern perspectives on sustainable development, as the next section documents. More interestingly, however, the Southern perspective on sustainable development may in the future *guide* developing economies along new paths toward prosperity which the North has not taken. From the evidence in this book and elsewhere (e.g., Munasinghe, 1993; Munasinghe and McNeely, 1994; Nagpal and Foltz, 1995), there is no reason to suppose that developing countries will follow the traditional development model of heavy industrialization leading to a post-industrial consumer society.

Is the notion of sustainable development applicable to the South? Paradoxically, it is currently being *applied* to the South by the international development banks (e.g., through the Global Environment Facility of the World Bank), but not much in the North, at least in proportion to the use of energy and materials. There are many possible reasons for this. One rather cynical reason might be a fear in the North of industrial competition from the South, which would necessitate maintaining agrarian economies in the South through the influence of appropriate development projects. That there is an *intention* on the part of the international development community to do this, or that that community is heavily influenced by political groups with this intent, is implausible. Another, and contrasting, reason might be the same optimism trap that the expansionists fall into with regard to the North's technical ability to become sustainable. This is the rather arrogant belief that the North could become environmentally sustainable within the time it would take to teach the South how to do it. That the international development community is occasionally capable of blithe faith in the North's technological capacities, as well as its own expertise in fostering sustainable economic development around the world, is entirely plausible. Whatever the reason, however, the imposition of the "new paradigm" of sustainable development has led, in part, to charges of economic colonialism and a partial or complete rejection of the concept in the South (see Chapters 2 and 4). In short, the Northern view of sustainable development is certainly informing the Southern view, but not always in ways the North would like.

The Northern perspective has dominated efforts to institute sustainability mainly because these efforts have focused on industrialized economies (most notably in the Netherlands) and on international projects in developing countries. Southern perspectives have not dominated efforts to institute sustainability because (1) the Southern voice is not heard as often as the Northern one; (2) most people in the South use natural resources at a relatively low intensity, and so are closer to sustainability in some sense anyway; and (3) developing country governments often do not see fit to constrain the improvement of their people's living standards—by whatever means necessary—while the North fails to enact curbs on its own profligacy.

If the Northern perspective on sustainable development is not easily applicable to the South, is it applicable to the North itself? The answer from the present analysis is a qualified yes, but the Northern perspective is incomplete, and a better job could be done if the North took a few hints from the South. Sustainability is not a homogeneous state, nor is there a single path toward it. The requirements for sustainability vary by developmental state, and the best way to meet these requirements is to share information among regions.

Southern Perspectives on Sustainable Development

Playing "Devil's Advocate" from Two Northern Standpoints

The most striking difference between the Southern and Northern perspectives on sustainable development emerging from this book is that the rampant self-interest which has historically characterized economic development in rich countries must be balanced with a stronger sense of community in order for global environmental sustainability to have a chance. A number of authors, especially Cavalcanti (Chapter 2), point to the lifestyles of indigenous peoples as models of environmental sustainability, subservience of individual to community interests, and non-material well-being. The implication of their arguments is that people in industrialized societies cannot achieve all of these things at once— equally because of the environment in which they live and because they have neither the will nor desire for it. Cavalcanti goes further by emphasizing that "not only is sustainability a requirement of the new concept of development, it is also a general prerequisite of life." On an evolutionary time scale, this is always true because life persists. The logical interpretation is, then, that under a business-as-usual scenario, with no interference, an indigenous lifestyle should outlast a 20th century industrialized one. The tragedy is that if the currently dominant, industrialized model fails as a result of overshooting natural limits for too long, it may take the indigenous one with it.

The Northern response might be: "What can we do?" Kiew (Chapter 4) is disparaging of such words because, to date, they have not been matched by actions. He states that "under the current accounting system, environmental sustainability is meaningless without profit in the financial world which governs the world's economy." The expansionist reply might be that financial profit is in our collective interest anyway because greater wealth buys greater environmental protection. This point is not as clear-cut as it seems because of the per-capita costs of that protection, but such an issue is beyond the present scope. At least, according to the expansionist argument, any international imbalance between the distribution of environmental harm and the costs borne for that harm should even out in the long term as the rest of the world approaches a Northern standard of living through economic growth.

Kiew's comment also alludes to the earlier point about lifestyles: the current accounting system values only profit and nothing else. The Northern expansionist might respond by arguing that a financial measure is preferred because the challenge for development essentially reduces to a valuation question on commodities. The Northern precautionist might invoke a scheme rather like the "biodiversity constraint" of Perrings (1991) or Smith (1996a), but one is still left wondering what form of development the "Southern" perspective desires.

There are, of course, many "Southern" perspectives. Perhaps the greatest misconception that people from industrialized nations could have is that people from poor countries speak with one voice. The multifarious views of the "South" are evident in many contexts in this book, but perhaps most strikingly in relation to the concept of "development" itself. Cavalcanti essentially advocates that communities not be required to "develop" as such, especially when they already live in relative harmony with nature. If environmental sustainability is the goal, then indigenous people are in fact the *most* fully developed along that path. In a sense, Cavalcanti's position can be envisaged as lying at one end of a spectrum of opinion regarding the value of "development," however defined. Other authors in the book sit at different points along this spectrum. Prakash and Gupta (Chapter 3) argue that however human activities change over time, they should be governed by institutional arrangements that are ecologically sustainable. Further, Chopra (Chapter 6) and Toledo (Chapter 11) each show how small-scale harvesting can link to the cash economy and be environmentally sustainable. Many of the other authors address "development" as a supposedly well-defined concept and analyze ways to make it sustainable. Ashley and Barnes (Chapter 8) encapsulate the opposite end of the spectrum from Cavalcanti by stating that "the development process has an important objective: improved livelihoods and opportunities for the historically marginalized poor." There is no right or wrong here—just a set of options whose relative potentials will depend on the specific circumstances.

Are these concepts of development what the people of developing countries want? According to Kiew (personal communication):

> The concept of being "developed" is a point of view. The people in the South and the poorer countries are given the impression by the media and their governments that the North is more developed. I personally do not subscribe to this. Otherwise I would have emigrated overseas to some country in the North.
>
> "Develop in what?" is the question to ask. The Penan in Borneo are developed in their life in a rain forest environment. They are

considered undeveloped for city life, a life of industries, trade, and commerce. Judging them in relation to an alien environment makes them *appear* undeveloped.

The North is considered developed due to the aspirations of the people in the South who have not attained these aspirations in terms of material products, technology, and energy use. It is also perceived that the living standards in the North are higher than in the South. People in the North, with their cars and other modern transport facilities, have more and better options in many things they do in life. The poorer people in the South can walk, cycle, or take a boat as their transport option. However, they would love to be able to drive, fly, and surf the Internet.

Kiew asserts in Chapter 9 that economic *growth* is what people in poor countries want: not so much an increase in their country's gross domestic product as an improvement in lifestyle, of which greater buying power is an important part. Moreover, many authors in this book argue that people in poor countries want, for the most part, to increase their well-being sustainably and that the North, by economic and environmental "colonialism," is preventing them from doing so. Moreover, people in the North are enjoying a high standard of living in an environmentally *unsustainable* way.

The potential for economic colonialism may exist not only in the agenda of the international development organizations, as already discussed, but is actually manifested in the international support of economic incentives denying people in poor countries equal competition in international markets. A common side effect is, of course, the degradation of the natural environment (e.g., Southgate, 1995). Yet people and organizations from the North routinely admonish developing countries for not conserving their natural resources, and some even take actions to obstruct the exploitation of those resources. The Northern environmentalist might argue that this so-called "environmental colonialism" is in *everyone's* best interests because poor countries could and should learn from the North's past mistakes. The Northern argument goes further: since the opinion in poor countries is that the North should clean up *its* act before telling others to clean up theirs, then, by the same token, developing countries should clean up their internal institutional and political organizations, because those changes alone would remove many of the economic perversities that obstruct sustainable development.

Issues of Concern

How do rich and poor regions differ with regard to the perceived issues of concern for individuals? In the rich countries, when the news media cover environmental issues, they tend to focus on dramatic natural events such as hurricanes, toxic algal blooms, and heat waves. The loss of biodiversity is featured occasionally, but poverty, or the inequitable distribution of wealth among people in different parts of the world, hardly ever appears in the media, except when a region is hit by famine.

This pattern is an example of geographical discounting (Hannon, 1994). The news media naturally cover stories that for the most part affect the people they serve. Global warming is an issue with many facets because it is a phenomenon with multifarious consequences, some of which impinge on the lives of people in industrialized regions. People in the North tend to be concerned about systemic problems in the natural environment because those are the problems that are perceived as threatening. Poverty is not a threat; it is mostly out of sight and therefore out of mind. By contrast, the clear message from this book is that people in poor countries care comparatively little about global warming and other systemic problems because they perceive the improvement of their lifestyles as a more pressing need. This is *temporal* discounting. Recall the earlier biological argument explaining why ecopopulism emerged in wealthy countries rather than poor ones.

The chapters in this volume find specifically that the greatest concerns of people in poor countries are food security, water security, health, education, land tenure, access to markets for their goods, and access to consumer goods, in approximately that order. By contrast, people in the industrialized world are mostly concerned about maintaining their standard of living, which is perceived as coming under attack from competition with emerging economies as well as unpredictable changes in global climate, the depletion of the ozone layer, and, increasingly in the 1990s, the spread of disease. Most indigenous people are, not surprisingly, concerned about preserving their way of life, which does not naturally produce the impacts that the rest of us worry about.

It is no shock to discover that some of the authors in this book argue for the North to do more to help the South develop economically, not by providing aid or loans but by removing barriers to trade and technology. The chapters in this book contain many creative ideas for building wealth without wrecking the natural environment. The problem that the South sees is that the North does not have the motivation to facilitate the successful application of those ideas. As Aguirre (Chapter 5) and others note, many sustainable practices that would generate wealth in the South require economies of scale that current institutional structures would not facilitate. For example, tropical forests valued economically rather than financially would appear a good investment in their intact state because they offer more than just timber (see Chapter 5), but could the sustainable extraction of that timber in one place compete with unsustainable practices elsewhere? Perhaps it could, if demand were managed appropriately, but the bodies that control international trade are slow to appreciate the potential in the link between trading patterns and sustainable resource use. If the World Trade Organization or other international organizations were ever to initiate a General Agreement on Trade and the Environment (as envisioned by DeBellevue et al. [1995] and Smith [1996a]), or an economic equivalent of CITES* to regulate the harvesting, production, and pricing of timber (Kiew, personal communication), then such initiatives would require the commitment of the rich countries from the start if they were to have a hope of succeeding worldwide.

Another perceived market distortion from the precautionist point of view is the existence of positive discount rates. Chopra, in Chapter 6, raises the well-known point that markets often do not account for the difference between individuals and society in their valuation of preserving resources for the future. The most widely cited culprit for this market imperfection is the social discount rate, which, it is argued, must be lowered or dispensed with altogether. However, Chopra points out that this type of action would "result in distortions in other investment decisions as well." She argues that "allowing for a cost of present use is a better method of adjustment" and supports this argument with an analysis of the harvest of non-timber forest products. Chopra's argument, which echoes Markyanda and Pearce (1991), is that altering discount rates is precisely the kind of crude policy that will fail to provide the right incentives for environmentally sustainable development in any part of the world.

Lessons for the North

Notwithstanding these international distortions, the prevailing impression from this book is that people in developing countries are pressing ahead with the sustainable creation of wealth with or without external assistance. Chapters 5 to 8 cover financial–economic issues, mainly focused around new ways to value and use natural resources. In addition to the work of Aguirre and Chopra previously discussed, Capistrano et al. (Chapter 7) analyze the prospects for sustainable fisheries in Bangladesh, and Ashley and Barnes (Chapter 8) show in a Namibian study that wildlife tourism is the most profitable and equitable use of land, provided all groups have a stake in the industry.

Chapters 9 to 12 deal with institutional, political, and grass-roots issues. Barrantes (Chapter 9) and Adger (Chapter 10) each show the importance of establishing rights to land, because these rights tie the long-term interests of the landholders to the long-term stability of the land's natural resources. Toledo (Chapter 11) gives an interesting account of how indigenous and peasant communities in Mexico can and are taking their long-established sustainable practices to the market, thus subverting the traditional process of agricultural mod-

---

* Convention on International Trade in Endangered Species.

ernization which has, at best, a mixed track record. Finally, Abdalla (Chapter 12) tackles the thorny problem of encouraging economic growth through increased energy use while promoting environmental sustainability. Note that economic *growth* is viewed as desirable, in contrast with the entreaties of Northern precautionists.

Many of the authors in this book strongly encourage community-empowered development as the most effective way to achieve economic and environmental goals. The international development programs of the post-war period are viewed for the most part as how not to encourage development. In Kiew's words (personal communication):

> Community-empowered development tends to be more sensitive to the needs of the environment [and] the people. Often in the case of development, the process is initiated from a remote area by a party with no appreciation of the environment to be altered. Colonialism, neo-colonialism or urban-colonialism are all the same in that they lack sensitivity and often are unable to respond to changes in the plan they have initiated. Such management processes lack the ability for appropriate, timely responses to ensure the best of a development project.

In summary, the mix of Southern perspectives on sustainable development de-emphasizes "development" as a centrally planned process and emphasizes community empowerment. Sustainability is seen as a natural conjunct to this empowerment, because the users of resources are responsible for their maintenance. However, remote powers still have a responsibility to provide institutional and political arrangements that facilitate the creation of wealth from intact natural systems rather than from the liquidation of those systems to meet short-term needs. This view is quite distinct from the technocratic "control" mentality that has dominated the international development agencies during the post-war period (see Norgaard, 1994). The message from the South is to let communities develop by themselves but provide them with a level playing field in the international marketplace.

One consequence of this view of sustainable development is that economic growth is often desirable. In fact, the liquidation of some natural capital in order to provide seed financial capital for other projects may also be desirable. The important difference between this process and the elimination of local ecosystems for single industries (e.g., felling rain forest for cattle ranching) is that it takes a long-term view. The ultimate goal under this scheme is, in Cavalcanti's words, to live "within the limits of the possible." One of the first requirements of this goal is the elimination of poverty, a process that usually necessitates economic growth. Cavalcanti (Chapter 2) examines these points in more detail.

Where the Southern and Northern perspectives meet is on distributional equity and the conservationist ethic. The former is seen as a necessary prerequisite for poverty alleviation and the latter is argued to be an attitude whose prevalence in the South is underappreciated in the North. In light of these comparisons, there may be value in a two-way transfer of "technology" between North and South. This would entail not only the transfer of "manipulative" technologies (scientific methods, machines, expertise, etc.) from the North to the South but also the transfer of "intellectual" technologies from South to North. Intellectual technologies would facilitate *thinking* about development in new ways (e.g., placing less emphasis on technocratic "quick fixes" and more emphasis on the power of communities to instigate sustainable practices).

A Synthetic Framework for Sustainable Development

It is clear from the foregoing discussion that the conception of sustainable development that grew out of the challenge to the neoclassical program in the North is not universally applicable, nor is the rather more heterogeneous set of ideas emanating from the South. The synthesis proposed here uses these somewhat differing perspectives to build a more complete vision of the requirements for the transition to, and maintenance of, sustainable economies. A synthetic framework for sustainability would necessarily be an evolving entity since the challenges and opportunities of sustainable development in the future will almost certainly be different from those today. It should be broad enough to frame a wide variety of prescriptions, just as the synthetic theory of natural selection frames a wide range of hypotheses in ecology and evolution.

In fact, the field of ecology itself provides a model for making sustainability operational. Ecology is an empirical discipline: almost all theoretical advancements have come from empirical research and, as a result, the field is composed of a heterogeneous collection of theories that apply to different phenomena at different scales. Unlike economics, where field research is not the main tradition, ecology does not possess a central theoretical core. This is in some ways its strength; like the disaggregated webs of interactions that ecologists study, ecological theory itself forms a web, albeit an incomplete one. The lesson for sustainable development and for ecological economists is twofold. First, we must use field research to guide policy recommendations; sustainability will not be made operational from inside an office. Second, do not be afraid of a disaggregated theory; there is no reason to suppose the existence of a "unifying theory," and searching for one is probably a waste of time. The research of the authors in this book demonstrates this lesson admirably.

If ecological economics, as the science of sustainability, should be a disaggregated theory, what would a synthetic framework for sustainability look like? Given that the goal is biophysical sustainability, any research directed toward that goal should first identify the relevant biophysical constraints and their values and then find ways to maximize social welfare (i.e., financial and non-financial wealth, distributional equity) within these biophysical constraints. In the language of neoclassical economics, this is roughly equivalent to maximizing social welfare subject to constraints, where the primary and overriding constraint is biophysical (i.e., this constraint must be satisfied before all others).

Just as the neo-Darwinian synthesis informs studies of ecology and evolution in almost all contexts, a synthetic framework for sustainability should inform the attainment of the sustainable goal in almost all circumstances. The difference between pure science and ecological economics is that research in ecological economics is issue driven, and therefore the components of a synthetic framework will be prescriptive rather than descriptive or explanatory. However, the essential structure is the same: Given a set of empirical observations, what explanations or prescriptions may be possible? For example, an ecologist might observe apparently altruistic behavior by an individual of one species toward an individual of another. The synthetic theory of evolution would view this as a paradox, which would motivate the researcher to investigate the system further for evidence that the "altruistic" individual is actually benefiting from its actions in terms of increased fitness. In the same way, a development economist might observe that institutional arrangements in a particular region to foster environmental conservation do not have the desired effect, and the people making direct use of environmental services are in fact causing environmental degradation, contravening the goal of environmental sustainability. This paradox should motivate further investigation to discover the source of this disconnect, which might, for example, lie in a lack of communication of the institutional arrangements or their conflict with local customs.

Following the parallel with the neo-Darwinian synthesis further, we can ask what criteria may be important for forming sustainable prescriptions. In ecology and evolution, the criteria that frame a question and permit the formulation of explanatory hypotheses include spatiotemporal scale, biological fitness, genetic polymorphism, primary productivity, social structure, trophic level, and so on. For sustainability, a synthetic framework may make use of:

- Spatiotemporal scale of investigation
- Ecological integrity (as measured by nutrient fluxes, biodiversity, or population sizes)
- Resource intensity of the human system and "balance of accounts" with the natural system
- Perceptions of the natural system by people at the primary level of interaction with it

- Perceptions of people far removed from the natural system
- Cultural norms and customs
- Governmental policies and attitudes within administrations
- Institutional structures
- Economic policies and operation of economic instruments
- Efficacy of communication among sectors of society about policies, institutions, and economic instruments, in terms of both dissipation of information to the populace and feedback to administrations
- Degree of centralization of power (level of community empowerment)
- Degree of corruption in society
- Distribution of wealth and welfare
- Desires of people with regard to wealth and welfare
- People's visions of sustainability

This list is really only a sample of a larger list. The idea is not to produce a complete list but rather to give a sense of the kinds of criteria a synthetic framework might contain. Two important principles emerge. First, this or any other list covers criteria for predicting and prescribing sustainability in both the "Northern" and "Southern" conceptions, and thus is integrative. Second, it is not necessary that every criterion in this synthetic framework be quantifiable. Ecologists, for example, have good reason to be suspicious of attempts by environmental economists to place financial values on species, and anthropologists, ethnographers, and social psychologists have every right to turn up their noses at attempts to quantify the criteria they study. The important task is to identify a set of rules of thumb which when applied consistently should produce approximately consistent outcomes. The interdisciplinary requirements of predicting and measuring sustainability are clear from the breadth of these criteria. As in the employment of the neo-Darwinian synthesis in biology, though, the fundamental concept is very simple: if we characterize the state of a system, what does the synthetic framework, itself derived from a broad empirical base, tell us about that state?

The conception of sustainability offered here integrates the full range of perspectives described earlier. Sustainability would:

- Be biophysically based
- Permit economic growth, in extreme cases through the sacrifice of some natural capital
- Foster distributional equity
- Measure wealth multidimensionally, not just in terms of money
- Promote conservationist ethics
- Empower communities
- Increase efficiency of resource throughput

- Spur the design of novel economic instruments
- Promote just and fair institutions, economic instruments, and business processes

A synthetic framework for sustainability would use the predictive criteria in the first list to make quantitative and qualitative measurements against the principles in the second list. Since environmental sustainability is probably a heterogeneous goal across locations and timescales, the employment of a framework such as this would also be heterogeneous. Like an expert diagnosis system, if we input the "symptoms" of a case, then the framework should help us guide our diagnosis and propose a remedy. Both the diagnosis and remedy would necessarily be multidimensional.

As a result of the neoclassical economic program in the North, a profound monetaration of values (Gowdy and McDaniel, 1995) has come to pervade the field of economics and society at large. It has also partially overrun other value systems among people in the South through international development projects founded on the maximization of income. A synthetic framework for sustainable development would reveal the many dimensions of "value," instead of just the financial dimension, and would promote the development of multi-attribute methods to measure value (Cavalcanti, Chapter 2).* A multi-attribute approach to sustainability delivers what is needed, namely, "rough answers to the right questions, rather than accurate answers to the wrong ones," as population biologist Robert May puts it (R. May, personal communication).

In the spirit of May's dictum, let us observe how a synthetic framework for sustainability might be applied to a real problem. A problem that has vexed economists and resource managers for years is how to harvest fish sustainably, that is, to maintain a fish stock in an ecologically stable state while exploiting it for a continuous stream of revenue (see Roughgarden and Smith, 1996). The ecological–economic solution is to solve first for ecological stability and second for the maximization of economic return and other metrics of social welfare. From ecological theory, the first part is relatively straightforward in that a regulatory body should simply set the combination of stock size and harvest rate to place the population in a stable zone, which is known (Roughgarden and Smith, 1996). The second part may be achieved by using an economic instrument, such as a tax, that varies directly with the size of the natural stock. As the stock

---

* Not all values can be measured quantitatively, for example, the aesthetic value of a wilderness or the psychological value of being self-employed (see Alonzo Smith, 1980). The challenge for multi-attribute systems is to be consistent with one another when measuring relative value. The fuzziness of multi-attribute systems will not appeal to neoclassical economists, who prefer clean answers, but it is this very fuzziness that is important in capturing the breadth of values that exist in reality.

declines through overfishing, the tax goes up, and before the stock decreases to an unstable size, fishing becomes uneconomic. This solution harnesses human self-interest* to ensure ecological and economic sustainability. More sophisticated instruments, such as stock-dependent options (Smith, 1996b), might do an even better job and be preferable for implementation. The important point is that there does not have to be an exact solution, a perfect optimum, at any point in time, but rather what May et al. (1978) call a "robustly self-correcting mechanism" that optimizes revenue over time.

Once the mechanics of the system are understood at the multispecies level (which is a more complex version of the single-species case [Smith, 1996a]), the next step is to create the institutional and political arrangements to regulate these mechanics. In the case of a marine fishery, the job is facilitated if ultimate control of the total allowable catch rests with a single regulatory body. If the economic instrument for regulation does its job well, fishing communities could be empowered to make their own management decisions while the regulatory body measures the stock, determines the available number of shares in the stock, and enforces against illegal practices (e.g., highgrading or using small mesh sizes). This rearrangement of institutional power would probably also encourage conservationist attitudes among the fishers themselves, and increase the efficiency of resource use because the fishers would themselves have a direct stake in the fish stock. By setting the ecological bounds at the outset, it is also possible to estimate how large the fishery can safely be. Therefore, setting ecological limits would not necessarily preclude economic growth.

A Heuristic for Sustainable Development

This last point has been the source of much confusion among economists and natural scientists during the 30 years that the debate on economic growth and environmental carrying capacity has remained in the policy arena. Understanding the relationship between ecological limits and economic growth is central to the institution of sustainability, but such an understanding is not always apparent. Some of the opinions of the authors in this book on economic growth are, on first encounter, a little difficult for a precautionist Northern reader to swallow. Kiew (Chapter 4) and others are in favor not so much of sustainable development in the South but sustainable economies and sustainable economic

---

* It is argued that self-interest is a characteristic common to all organisms and is irreducible. Economic instruments for sustainability will probably be most powerful if they harness this self-interest. The assumption of self-interest does not preclude cooperation or altruism, but these phenomena are hypothesized to be explicable by it.

growth—a strictly expansionist interpretation of sustainability. It is instructive to show on a very simple level that this expansionist interpretation is, in fact, compatible with a precautionist one under certain conditions. To do so, a heuristic is introduced which may help to guide questions within the synthetic framework proposed above.

This heuristic shares some similarity with the *IPAT* identity of Ehrlich and Holdren (1971). In the present case, however, we are interested not in environmental impacts (the *I* in *IPAT*) but instead in the conditions for sustainability. The heuristic considers the size of the economy (E), population (P), technology (T), trade (M), and the throughput of resources (R). Clearly, the size of the economy is a function of the other four parameters:

$$E = f(P, T, M, R) \tag{1.1}$$

where P, T, M, and R are assumed to be multiplicative.*

If the throughput of resources, R, is increasing, then, *ceteris paribus*, E will also increase. In the world today, we observe that P, T, M, and R are all increasing and that the increases in P, M, and R combined outweigh any offsetting increases in T. Improvements in T may or may not be environmentally beneficial, but in cases where they are, they are assumed to increase the efficiency of resource throughput, thus reducing R per unit of P.

For sustainability, the objective is to maintain R steadily at a non-damaging level. What are the requirements for this? The other factors in the heuristic are somewhat interdependent and also dependent on external factors not included in it; for example, trade (M) is not only a function of population and technology but also of institutional and legal factors that may facilitate or hinder it. Thus, P and T could be static but M might still be increasing, and therefore so would E. More interesting is that *all four* factors could be increasing but R would still remain steady if the improvements in T were *environmentally beneficial.* This observation drives home the point that zero economic growth may not be a *necessary* requirement for environmental sustainability. What may be true in the case of a fishery may, therefore, be true generally.

Of course, we have so far neglected the issue of carrying capacity. The economy and population cannot grow indefinitely, although there is considerable uncertainty about their biophysical ceilings (e.g., Simon, 1986; Daily and Ehrlich, 1992; Moffat, 1996). At any rate, as long as the carrying capacity is not exceeded, P and E may increase and R could remain steady and non-damaging as long as the aforementioned requirement concerning T is satisfied. A plausible scenario for environmental sustainability might be a static population in con-

---

* Alternatively, one could frame the heuristic in terms of the per-capita size of the economy: $E/P = f[(T, M, R)/P]$.

junction with economic growth brought about by trade in environmentally benign technologies. In this scenario, economic growth may be beneficial for sustainability since it would fuel the requisite technological development and refinement. It is reasonable to hypothesize that if economic development took the appropriate set of paths, this state could be reached in a few generations. Of course, the requirements for sustainability at that point would be particularly stringent since we would be paying for past excesses in terms of economic potential foregone by environmental degradation. Only over a much longer period hence would these requirements relax.

The foregoing analysis argues that the attention on population and economic growth in the academic literature and the popular press is perhaps misplaced. It is not so much how many of us there are, nor how fast the economy is growing, but *what we do*. If environmental sustainability is to be reached, then attitudes must change with respect to technology. The North in particular has barely started to address, on environmental grounds, why any given industry should exist in the first place or how it should be structured to comply with biophysical constraints. Southern admonishments to the North to put its own house in order before criticizing environmental performance in developing countries are, in light of this analysis, well placed.

On the face of it, there appears to be some inconsistency between the insights of this heuristic and the concept of a steady-state economy proposed by Daly (1991). According to Daly, a steady state entails a constant stock of physical capital and a constant population. This necessitates low rates of resource throughput, which in turn requires lowering production and consumption rates as well as birth and death rates in the population. He emphasizes that sustainable development "implies a different direction of technological progress [compared with today's] that squeezes more service per unit of resource" (Daly, 1991, p. 253), thus maximizing resource-related efficiency rather than economic efficiency and maximizing productivity. Certainly there is overlap between Daly's prescription for sustainability and the prescriptions in the present volume, but some authors' emphasis on the occasional *necessity* of population and economic growth lies outside Daly's conception. It seems that the steady-state economy as envisioned by Daly represents a subset of possibilities for dynamic stasis within a broader set that would permit economic growth in technologies which conserve or enhance environmental integrity. It is reasonable to expect that at least while populations are increasing, economies will grow. The challenge is to make them grow the right way. Eventually, in a world which contains a non-growing population, we may attain a steady state, but for the foreseeable future the distributional inequities across the globe will represent a powerful force for increases in wealth.

Heuristics such as the one presented here may be able to identify other allowable conditions for sustainability. For example, if some natural capital is liquidated, *R* is effectively increased because natural resources are poured into the economy *and* the size of the natural sink for human wastes is reduced. Trade would presumably also increase and the economy grew. Since the objective is to maintain natural capital, this action is grossly unsustainable and should be discouraged except under critical circumstances.

Conclusions

Synthetic frameworks for sustainable development, guided by simple but informative heuristics, are valuable for directing attention to the important drivers for sustainable development, but they do not inform the mechanics of development itself. For that, we need much more detailed, customized plans, which is what the rest of this book is about. It is important to note that the prescriptions in the present volume originate from work in developing countries for three reasons:

1. The developing country perspective has been neglected in the literature.
2. Most economic growth and development in the next century will take place in today's poor countries, and this book illuminates alternatives to the development path taken in the industrialized nations of the North.
3. The ideas, analyses, and prescriptions contained herein can inform the attainment of environmental sustainability in the rich countries as well as the poor.*

Many of the steps toward sustainability described in this book will involve economic growth (i.e., an increase in the volume of goods and services traded), and this, according to the authors of the book, is a desirable thing. It also need not contravene natural constraints if other conditions are satisfied, as the heuristic presented above suggests. The practical solutions that follow point toward methods for bringing society's precautionist and expansionist tendencies together into a kind of balance.

As much reference guides for planning as academic texts, the following chapters describe approaches that appear to work, approaches to avoid, and how development programs may be adapted to different circumstances or different

---

* Some work on practical approaches to, and applications of, sustainability already exists, most notably van den Bergh and van der Straaten (1994), but these are mostly assessments of existing sustainable development policies and do not draw specifically from the broad range of developing country perspectives.

regions. In addition, the foregoing discussion and the findings that follow may encourage thinking about sustainability in a more integrative way than to date.

Numerous signposts exist which policymakers could use to guide their decisions, if their intentions are for sustainability. For example:

- Are the natural dynamics and constraints for any given system known? Can they be known, or can the uncertainty be managed?
- What lifestyles do people want?
- What practices are currently disruptive, environmentally or socially? Can they be adjusted or replaced?
- How much can real decision-making power be devolved to communities? Are there feedback mechanisms in place to pass information from scattered communities to central authorities?
- Which is preferable in any instance: low-tech or high-tech?
- How can people's innate self-interest be harnessed for sustainability?
- As a decision maker, are my decisions motivated by my own self-interest or the broader goals?

Most important is a continual awareness that technological improvements to serve only the ends of consumption, without any recourse to the sources of that consumption, are grossly unsustainable. In this book, we try to find ways to steer technology and consumption in directions that generate the conditions for long-term stability, namely, adequate economic welfare combined with environmental integrity.

Just as indigenous people can show the rest of us what this concept means by living the very distillation of sustainability, the practical approaches described in this book show us how to adapt our economies to emulate that existence. These steps are small and may, in a sustainable future many generations hence, appear weak-hearted, but, as an old Chinese proverb says, even a long journey must begin with a small step.

Acknowledgment

The author wishes to thank L. Brien for valuable comments on an earlier draft.

References

Alonzo Smith, G. (1980) The teleological view of wealth: a historical perspective. In: *Economics, Ecology, Ethics: Essays Towards a Steady-State Economy.* H.E. Daly and K.N. Townsend, Eds. W.H. Freeman, New York, NY, pp. 215–237.

Boulding, K. (1966) The economics of the coming spaceship Earth. In: *Environmental Quality in a Growing Economy.* H. Jarret, Ed. Johns Hopkins Press, Baltimore, MD, pp. 3–14.

Costanza, R. and Patten, B.C. (1995) Defining and predicting sustainability. *Ecological Economics,* 15: 193–196.

Daily, G.C. and Ehrlich, P.R. (1992) Population sustainability and Earth's carrying capacity. *Bioscience,* 42: 761–771.

Daly, H.E. (1968) On economics as a life science. *Journal of Political Economy,* 76: 392–406.

Daly, H.E. (1972) *Toward a Steady-State Economy.* Freeman, San Francisco, CA.

Daly, H.E. (1991) *Steady-State Economics,* second edition. Island Press, Washington, D.C.

DeBellevue, E.B. et al. (1994) The North American Free Trade Agreement: an ecological–economic synthesis for the United States and Mexico. *Ecological Economics,* 9: 53–72.

Ehrlich, P.R. and Holdren, J.P. (1971) Impact of population growth. *Science,* 171: 1212–1217.

Georgescu-Roegen, N. (1971) *The Entropy Law and the Economic Process.* Harvard University Press, Cambridge, MA.

Gowdy, J. and McDaniel, C.M. (1995) One world, one experiment: addressing the biodiversity–economics conflict. *Ecological Economics,* 15: 181–192.

Gowdy, J. and O'Hara, S. (1995) *Economic Theory for Environmentalists.* St. Lucie Press, Delray Beach, FL.

Hannon, B. (1994) Sense of place: geographic discounting by people, animals and plants. *Ecological Economics,* 10: 157–174.

Kaufmann, R.K. and Cleveland, C.J. (1995) Measuring sustainability: needed—an interdisciplinary approach to an interdisciplinary concept. *Ecological Economics,* 15: 109–112.

Markyanda, A. and Pearce, D.W. (1991) Development, the environment and the social rate of discount. *The World Bank Research Observer,* 6: 137–152.

Martinez-Alier, J. (1987) *Ecological Economics: Energy, Environment and Society.* Blackwell, Oxford, U.K.

May, R.M. et al. (1978) Exploiting natural populations in an uncertain world. *Mathematical Biosciences,* 42: 219–252.

Mayr, E. (1982) *The Growth of Biological Thought.* Princeton University Press, Princeton, NJ.

Moffat, A.S. (1996) Ecologists look at the big picture. *Science,* 273: 1490.

Munasinghe, M. (1993) Environmental Economics and Sustainable Development. World Bank Environment Paper No. 3. World Bank, Washington, D.C.

Munasinghe, M. and McNeely, J., Eds. (1994) *Protected Area Economics and Policy: Linking Conservation and Sustainable Development.* World Bank, Washington, D.C.

Nagpal, T. and Foltz, C., Eds. (1995) *Choosing Our Future: Visions of a Sustainable World.* World Resources Institute, Washington, D.C.

Norgaard, R.B. (1994) *Development Betrayed: The End of Progress and a Coevolutionary Revisioning of the Future.* Routledge and Co., New York, NY.

Perrings, C. (1991) Ecological sustainability and environmental control. *Structural Change and Economic Dynamics,* 2: 275–295.

Reid, D. (1995) *Sustainable Development: An Introductory Guide.* Earthscan Publications, London, U.K.

Roughgarden, J.R. and Smith, F.D.M. (1996) Why fisheries collapse and what to do about it. *Proceedings of the National Academy of Sciences,* 93: 5078–5083.

Simon, J.C. (1986) *Theory of Population and Economic Growth.* Blackwell, Oxford, U.K.

Smith, F.D.M. (1996a) Biological diversity, ecosystem stability and economic development. *Ecological Economics,* 16: 191–203.

Smith, F.D.M. (1996b) Options to harvest: applying financial option theory to the regulation of natural resource harvesting, paper presented at IV Biennial Meeting of the International Society for Ecological Economics, Boston, MA, August 4–7.

Southgate, D. (1995) Economic progress and habitat conservation in Latin America. In: *The Economics and Ecology of Biodiversity Decline: The Forces Driving Global Change.* T.M. Swanson, Eds. Cambridge University Press, Cambridge, U.K., pp. 91–98.

van den Bergh, J.C.J.M. and van der Straaten, J., Eds. (1994) *Toward Sustainable Development: Concepts, Methods and Policy.* Island Press, Washington, D.C.

WCED (1987) *Our Common Future.* World Commission on Environment and Development and Oxford University Press, Oxford, U.K., 400 pp.

Wilson, E.O. (1992) *The Diversity of Life.* Belknap, Cambridge, MA.

Philosophical Perspectives: Third World and First World Compared

Part I

Patterns of Sustainability in the Americas: The U.S. and Amerindian Lifestyles*

2

Clóvis Cavalcanti
Institute for Social Research, Fundação Joaquim Nabuco, Recife, Brazil

Abstract

Evidence from anthropology and ethnoscience has shown that a society can live within the limits of the possible and have a joyous life. This contrasts sharply with modern living standards, which are obtained at high ecological costs, without necessarily permitting people to combine material affluence with a good livelihood. This chapter uses evidence from studies done by anthropologists in Brazil to compare certain aspects of the indigenous people's way of life with some attributes of U.S. society. The comparison is undertaken with a view toward discovering those sustainable characteristics of the Indian lifestyle that are not observed in the United States. It is not intended to convey a picture of the Brazilian Indians as a countercultural utopia in the face of progress depicted as a straight-jacket, but instead calls attention to a form of knowledge that can be extremely important in devising a sustainable future for humanity. In fact, the Indians exhibit a harmonious way of life in terms of man–nature relationships and thermodynamic thrift. At the same time, they have always made a favorable impression on outsiders who have made contact with them.

---

* A previous version of this paper was presented in October 1994 at the III Biennial Meeting of the International Society for Ecological Economics in San José, Costa Rica.

1-57444-077-2/97/$0.00+$.50
© 1997 by CRC Press LLC

27

Introduction

Sustainable development, sustainability, and the like have been topics of much discussion in recent times. The reason is simple. It is not possible to expect that economic growth will continue to follow the same path of unlimited expansion that characterized the post-World War II experience. This pattern of development has not brought genuine progress; poverty has increased in many parts of the world, including sections of the rich countries. The idea that development can be self-sustained, as espoused in the works of many economists or economic historians, such as W.W. Rostow (1960), has been abandoned to the extent that the economy cannot be sustained by itself. New concepts have been elaborated concerning the constraints that the environment imposes on the economic process, not to mention the meaningful field of ethical constraints. Instead of merely referring to economic development as something that exists by itself, emphasis has thus been placed on sustainable development—development that is permitted by the rules of the ecosystem and which satisfies simultaneously the requirements of justice and social equity.

To take into account this new reality of sustainability, a new field of inquiry, ecological economics, has evolved with a focus on the need to reconcile material progress with sound management of the environment. This means that some sort of balance must be reached between the tendency toward the increasing demand for resources (caused by the expansion of the economy and by population growth) and the constancy of our ecosystem, or the invariance of the amount of matter and energy at our disposal. Such is the essence of the notion of sustainability, no matter how it is defined, either in theoretical or operational terms. The question is how to maintain the potential productivity of the system that supports not only the economy but life itself. It has now become evident that the paradigm of economic development—or growth, for that matter—which is practiced in the modern Westernized world does not lead to the necessary compatibility of the ecological base with the goals set for the economic system. However, the improvement of the living conditions of the extremely poor, a moral duty of society, is normally associated with the augmentation of aggregate product. Growth is deemed indispensable in order to eliminate or sensibly reduce extreme poverty. In this context, nature is simply asked to provide the resource layer that can sustain the unrestrained expansion of the economy. Almost no one questions the extent to which nature can fulfill this function attributed to it. Some even contend that material progress must not be stopped because our modern life-support system is composed of artifacts (see, for example, Simon, 1987).

This is the background against which we may juxtapose a different understanding of the problem. To begin with, we should accept the remark of Sachs

(1992) that "a global monoculture spreads like an oil slick over the entire planet," which leads us to admit the inevitability of economic development as preached since the start of the Cold War. To develop, therefore, one has to follow the guidelines established by the experience of the industrialized countries, which means allowing oneself to be sucked into the homogenizing pool of cultural traits (market, state, science, technology) peculiar to the Occident. This is what is expected from a world which actually embraces multifarious cultural elements and traditions, some of which are simply incompatible with the idea of growth. Furthermore, the modern understanding of truth, which is not, as pointed out by Faber et al. (1994a), the only understanding of truth that is possible, constitutes the sole framework of ideas adopted to rule decisions related to technological progress, economic performance, and social change. We are led to think that the options for a decent survival of people on earth are reduced to the paradigm offered by the first world, the countries of the Organization for Economic Co-operation and Development (OECD), and some exceptional cases of high economic achievement. However, the findings of researchers in the fields of anthropology and especially ethnobiology, although quite incipient, reveal indigenous perceptions about ecology and the utilization of natural resources (Posey, 1992a) which show that "there are options for the survival of Man in the Biosphere" (Posey, 1990). These options can be found in the lifestyles of native peoples and serve to caution against the tendency to promote economic development at such a pace that it cannot be halted in time to prevent the destruction—sometimes irreversible destruction—it is about to cause (see Posey, 1992a).

The approach to economic issues which suppose the existence of real ecological boundaries (i.e., that the planet is a non-growing entity, that matter and energy are conserved) is, in my contention, something that can be conducted with support of traditional knowledge and the practices of indigenous peoples, like some of those still found in Amazonia. In other words, the ecological treatment of economic problems grounded in modern Western thinking tends to elide important perspectives and to view nature from the Cartesian, less holistic perspective of dominance of it by humans. Traditional knowledge, now more and more widely accepted, offers sound alternatives for resource use and management based on experience and close monitoring of practices of native peoples over very long periods. It can supplement modern science and open new horizons of understanding. The solution of ecological–economic problems has much to learn from it, chiefly because to some extent it is the only source of alternative models of development that can be ecologically and socially sound (see Museu Goeldi, 1987).

The purpose of this chapter is to delve into the question of sustainability by employing evidence from anthropology and ethnoscience concerning Amerindian groups to help comprehend how a society can live within the limits of the possible and still have a joyous life. Support for this task is provided by the literature in anthropology, which is not specifically directed toward the study of sustainable features of given social groups. The problem is that the evidence is sparse, and there is no systematic way of showing how sustainability is achieved. On the other hand, as an economist, I am not well trained to deal with either the issues or the methods of study of anthropologists and ethnoscientists. I am aware, however, of the pitfalls that exist when we enter fields of inquiry other than our own. As a practitioner of ecological economics, I also think that we cannot avoid doing inter- and transdisciplinary work within this new area of study and that we should follow Georgescu-Roegen's (1971) advice: "Venturing into territories other than one's own" is a project "definitely worth undertaking." When he successfully undertook the analysis of the relationship between the second law of thermodynamics and the economic process, he built "on the writings of the consecrated authorities" in the field of physics, noting that "even so, one runs substantial risks" (Georgescu-Roegen, 1971). I face the same risks and challenges in invading anthropology and ethnoscience, but I find the task worth carrying out, because I find that economists in general have much to learn from ethnology.

The findings presented here are the result of research mainly in the Amazon with Amerindian tribes still living there, whose lifestyles are worth examining. It is not the intent to convey a picture of the Brazilian Indians as a countercultural utopia in the face of progress depicted as a straightjacket, but rather to call attention to a form of knowledge that can be extremely helpful in devising a sustainable future for humanity. The Indians exhibit a harmonious way of life in terms of man–nature relationships and thermodynamic thrift. As such, they deserve much more attention than they have been given up to now, although some—supposedly not in the academic world—still view them as having an inferior culture (as a Brazilian authority claimed about the Yanomami Indians in the late 1980s), perhaps as the remains of the belief that the Amerindians were not human beings, which was held until Pope Paul III (1534–49) in his bull, *Sublimus Deus,* removed such a label.

Paradigms of Sustainability

The Economist regularly publishes a chart of risk ratings by country for 26 "emerging markets." The chart shows a summary of national credit risk indices based on strictly economic and political factors. What if those countries were charted in terms of global environmental soundness or ecological sustainability

Figure 2.1 **Paradigms of sustainability measured in terms of a hypothetical scale of degrees of sustainable lifestyles.**

or if countries were listed according to the latter parameter by ascribing a rating to different lifestyles? I have been trying to compare two very different life paradigms in relation to this point (Cavalcanti, 1992). One paradigm (Figure 2.1) is found in the United States, with its high rates of per-capita resource consumption. The other paradigm is the opposite of the former in terms of frugality: it is the lifestyle of the Brazilian Indians who, in wild state, still inhabit portions of the Amazon and whose consumption needs are satisfied with very austere standards. Sustainability is much higher in the case of the Indians who live within what we can call the limits of the possible without causing social or ecological stress. This situation seems to fulfill the condition of a steady-state economy (Daly, 1980) by slowing down the energy flow.

If ecological economics is the science and management of sustainability, then the Amerindian paradigm cannot be neglected in terms of what it teaches. It is precisely this point that is underlined by the distinguished Colombian-born anthropologist Gerardo Reichel-Dolmatoff (1990) when he says that "the Indians' way of life reveals to us the possibility of an *option*, of a separate strategy of cultural development," which provides, in his view, "alternatives on an intellectual level, on a philosophical level," or alternative cognitive models that "we should keep in mind." The Amazonian Indians copy the patterns of nature, assimilating the principles they observe in the natural ecosystems; their lifestyle thus reflects the basic systemic wisdom (Bateson, 1972) inherent in nature. The Indians' cosmovision is based upon their knowledge, whereas the Americans' relies on modern science (the scientific logos). Sustainability is observed by the natives of Amazonia insofar as they plan according to the needs of future generations and take care of the living conditions of other species, thus assuring the preservation of biodiversity. With a strong sense of community, the interests of the individual among the Indians are not pursued unrestrictedly, much in contrast to the American paradigm, where man–nature relationships are defined according to traditional Western thought from an anthropocentric standpoint.

Reichel-Dolmatoff (1976), referring to the Tukano people's world views, says that their cosmological myths "do not describe man's place in nature in terms of dominion, mastery over a subordinate environment." He also remarks that the primitive tribes of the Amazon Basin, which to some are "fossil societies" that do not have anything to teach us, are not incomplete in the sense that they have not evolved but rather have developed highly adaptive behavioral rules for survival "framed within effective institutional bodies" (Reichel-Dolmatoff, 1976). The set of ecological principles elaborated by the Indians is combined with a system of social and economic rules leading to "a viable equilibrium between the resources of the environment and the demands of society" (Reichel-Dolmatoff, 1976). It is worth noting here that Reichel-Dolmatoff, who studied the Tukano for over 50 years, found that there is little concern among them for maximizing short-term gains or for obtaining more food or raw materials than are actually needed. In the Indians' view, "man must bring himself into conformity with nature if he wants to exist as part of nature's unity, and must fit his demands to nature's availabilities" (Reichel-Dolmatoff, 1976; cf. B. Commoner in Tiezzi, 1988). This is simply the opposite of Aristotle's *pleonexia*, or the wish to have and to always have more, which is "the driving force of modern productive work" (see Faber et al., 1994). Other anthropologists have arrived at similar conclusions; Viveiros de Castro (1992), for example, alluded to the Araweté in Pará State in Brazil, whose first contacts with white people occurred only in the late 1970s and whose culture he found to be wholly strong, gay, original, and imaginative. The focus of Indians' interest is conservation of their territory. This was clearly expressed by a Yanomami tribesman in a letter to Brazil's President Sarney dated September 1, 1989: "Our thought is our land. Our interest is to preserve the land, not to create diseases for the people of Brazil, and not only for the Indians" (CCPY et al., 1990).

In contrast to modern perceptions, and the American paradigm as well, the dominion of concepts and fundamental aspects of our civilization by the Indians, such as money, ownership, the state, sexual taboos, division of labor, misery, domination, and so on, is extremely precarious (Viveiros de Castro, 1992). The Yanomami Indian's letter to President Sarney referred to above, explicitly: "We do not know anything about money, shoes, clothes....The government does not know our custom, our thought" (CCPY et al., 1990). When these elements of civilization are introduced into the Indian society, they provoke serious disturbances, as indicated by Betty Mindlin, who has studied tribes in Rondônia State. The results of her findings show that "the use of money modifies food habits, reduces the rhythm of agricultural work, causes undernourishment, not because of scarcity properly...but for a new utilization of time, new behaviours...[and]

money is not distributed with the same fairness, according to the village's laws of reciprocity. It prevails over kinship, over the previous rules for a good living: and our society knows well about it" (Mindlin, 1994). Similar disturbances following contact with white people tend to increase the inequality between Indian men and women (Mindlin, 1994).

Ecologically sound land-use planning is a common feature of Indian societies in Amazonia, although the natives might occasionally contribute to the degradation of their lands. The Indians, for instance, submit birth rates and harvest rates (the exploitation of the physical environment) to adaptive rules to ensure individual and collective survival and well-being (see Reichel-Dolmatoff, 1976). This task is conducted by the shamans, who manage resource use. Some measures traditionally undertaken by the Amazonian indigenous peoples, such as the protection of forests on the banks of a river as a resource for fish subsistence (which the Wanana have long practiced; see Chernela, 1989), have only recently been considered scientifically sound. All this is implemented with a sense of profound respect for nature, from which the Indians copy their methods of environmental management. Viveiros de Castro (1992), speaking of the Araweté, comments on the Indians' simple technology and high capacity for improvising. It is not surprising, then, to discover that local communities and tribal groups are "the most effective managers of the resources" available to them (Panayotou, 1991). They know how to live within the limits of the possible as well as how to take care of socially disruptive behavior (aggression in interpersonal relations), which the Tukano submit to rules that serve to counterbalance it (Reichel-Dolmatoff, 1976).

The legacy of centuries of balanced environmental management by native societies could be appreciated by the first Portuguese to come to Brazil in 1500 (see Cortesão, 1943). What they encountered was a magnificent, beautiful country (Ribeiro, 1987) with abundant vegetation, pristine water, and plentiful game and fruits (Gandavo, 1924); that same environment can be found today in parts of the Amazon. The primitive inhabitants of Brazil in 1500 were good-looking, healthy, and strong (Cortesão, 1943); these same attributes were noted by anthropologists who did research among Indians during this century. Viveiros de Castro (1992) pointed out that in 1981 the Araweté were "visibly well nourished." Seeger (1980) arrived at the same conclusions in relation to the Suyá, underscoring their adequate diet. Baldus (1970) commented that the Tapirapé (men and women, adults and children) were used to enduring long journeys of 40 to 50 km through the forest and savanna without becoming exhausted. This, he notes, is proof of the Indians' vitality, in spite of their short average life span. Writing in the 16th century, Gandavo (1924) made analogous remarks. The Tupinambá, according to Sousa (1971), also a 16th century writer, were excellent divers, swimmers, runners and rowers and showed a great ability to climb trees and to jump.

In addition to being strong and healthy, the Indians seemed to be very happy with their lifestyle. This is stressed, for example, by Viveiros de Castro (1992), who studied the Araweté for 14 years. In his words, "to live with the Araweté is a fascinating experience. Few human groups, I imagine, are so easy to deal with, so joyful in their daily life...absolute in giving and asking, unrefrained lovers of the pleasures of life." An equivalent state of affairs was found among the Tapirapé by Baldus, who in 1935 came upon a constant atmosphere of joy in the village where he lived. "All the environment is tenderness. No one yells at anyone and even the dogs which bark at me on my way are taught discreetly to respect me. Everywhere I find gladness and laugh" (Baldus, 1970). He adds in the same passage that "courteousness...manifested itself in various degrees as a general pattern of behaviour," concluding that the Tapirapé "were the most joyful people" he had found in his life. Bruce Albert, an anthropologist conducting research with the Yanomami, reproduced an interview with the Indian Davi Kopenawa after the invasion of the Yanomamis' territory, in which he says, "Now you tell the other white men...how we were, with good health...how we did not die easily, we did not have malaria. Tell how we were really happy. How we hunted, how we gave parties. You saw that...today the Yanomami do not build their big houses any more...they only live in small shanties in the woods, under plastic sheet. They do not even grow crops, they do not go hunting any more because they become ill all the time" (CCPY, 1990). This is in stark contrast to the still isolated Araweté, about whom Viveiros de Castro (1992) concluded, "This is not a desperate culturally demoralized people, composed of sick, alcoholic, hungry and fearful persons—up to now."

Other characteristics of the Indians who inhabited Brazil in 1500 and who inhabit Amazonia today suggest that they not only were well adapted to the environment, enjoying good health and a joyous life, but that they are peaceful and courageous (Baldus, 1970), that they do not accumulate anything (Gandavo, 1924), that they do not worry about locking up their belongings and are not familiar with stealing (Cardim, 1939), that they are hospitable, and that they have a strong sense of community, generosity, and communion (Sousa, 1971). Sousa (1971) sums up his observations by saying that the Indians he described, the Tupinambá, were apt to be Franciscan friars because of their propensity to give away possessions. The dissimilarity to a lifestyle conceived around the craving for all kinds of possessions is striking. We face here two very different perceptions of life, with serious implications in terms of environmental health and social equilibrium. It is no surprise, then, that Faber et al. (1994a) raise the following question: "Are not the increasing problems of social disorder, violence, drugs, etc. consequences of the level of our present standard of living?" I do not intend to resuscitate the Rousseau-esque myth of the *bon sauvage*, but indigenous experience deserves to be seriously considered by research on sustainable development, for the Indians are "a diligent, intelligent and practical people who have adapted successfully for thousands of years in the Amazon" (Posey, 1992b), making their livelihood in many different ways, according to the local constraints.

The Amazon Indians' System

Using evidence provided by anthropology and ethnoscience, some of the chief characteristics of Indian societies still living in Amazonia are listed in Table 2.1. It provides a summary of the sparse information found in the literature, where the subject of sustainability springs up mostly in an implicit, unsystematic way and is mixed with such topics as kinship, material culture, rituals, description of daily life, custom, traditions, myths, and so on. The picture offered by Table 2.1 contradicts the anthropological evidence offered by Lewis (1992) showing that pre-modern peoples did not live in harmony with their surroundings. It conforms, however, with the remark of anthropologist Berta Ribeiro (1987) that the Indians treat their surroundings with respect, love, and care to ensure the permanence of nature as a source of food, human welfare, and a cure for illnesses. It also reflects what ethnologist Darrell Posey and other researchers discovered in their important work at Belém's Goeldi Museum; the basic aspect of the natives' management of natural resources is "a long term perspective, with emphasis on preservation, and not on the destruction of native resources of Amazonia" (Museu Goeldi, 1987). It reflects, moreover, recent findings of prehistoric archaeology that the Indians' health conditions became much worse after the conquest by Europeans, demonstrating that living patterns deteriorated as a result of a subsequent lesser concern for the environment in the Amazon (Roosevelt, 1991). In fact, the knowledge that is gained from anthropological sources is that the beliefs and attitudes centered on life which the Indians exhibit, in combination with the hundreds of little things they do, think, or avoid, as well as their perception of the universe, "form a highly structured order" (Reichel-Dolmatoff, 1990). In the case of the Kayapó (*Mebêngôkre*), their knowledge constitutes an integrated system of beliefs and practices such that, for instance, "each and every *Mebêngôkre* believes that he or she has the ability to survive alone in the forest for an indefinite time" (Museu Goeldi, 1987).

One aspect of the Amazonian Indians' view of nature, noted in relation

Table 2.1   The Amazonian Indians' System: Chief Characteristics

Energetics
- Basic source of energy is the sun
- Extreme thermodynamic thrift
- No use of fossil fuels
- No shortage of energy

Economics
- Very clear man–nature relationship
- Sustainable and efficient use of natural resources, preserving productive ecosystems and environmental quality
- Scale of activities within the carrying capacity of their territory
- Daily consumption of materials remains constant over time
- Life supported by the biological products of photosynthesis (water, forests, clearings)
- No economic development (no growth, of course)
- Satisfaction of basic needs
- Ignorance of money and ownership
- No accumulation of wealth
- No income inequalities (idea of poverty ignored)
- Intergenerational equity
- Simple, "soft" technology
- Absence of technological improvements
- Itinerant agriculture
- No use of inorganic chemical products of any sort

Demographics
- Population held within given limits
- Dispersion
- Small villages, small population units

Culture, philosophy
- Nomadic
- Simple material culture
- Complete observation of the laws of nature
- Respect for biodiversity
- Nature not exploited, but revered
- Long-term perspective
- Holistic, integrated view of life, reality and environment
- Apparent enjoyment of life

especially to the Tukano, is its "remarkable semblance to modern systems analysis," according to Reichel-Dolmatoff (1976), who points out that the Tukano's ecological theory "conceives the world as a system in which the amount of energy output is directly related to the amount of input the system receives." Energy in such a scheme should never be used without being restored as soon as possible. The restitution to nature of the energy potential utilized involves complex rules, practices, and rituals "whose totality corresponds to a way of life, to an integrated system" (Reichel-Dolmatoff, 1990). This represents a sharp contrast to a lifestyle dependent on the ever-increasing consumption of goods and non-renewable energy sources.

It is worth noting that the Amazon Indians showed a very different geographical distribution before conquest, when human occupations of large portions of the region ("paramount chiefdoms") were established. With the arrival of the first colonizers, the natives were dislodged to soil-poor, inter-river forests of the Amazon Basin (Roosevelt, 1991). These soils are, in effect, some of the world's most nutrient poor (Posey, 1992a), but the Indians adapted their techniques for living in harmony with nature, obtaining favorable results without degrading or exhausting the environment (Posey, 1992a), a pattern of behavior still witnessed among present-day remnant groups. The paramount chiefdoms that existed in Amazonia developed intensive food production, urban-scale settlements and monumental earth construction, "including the earliest pottery-age cultures in the hemisphere" (Roosevelt, 1991). Dispersion and the formation of smaller communities occur after the 16th century. This finding of recent archaeological work reaffirms the enormous ability of the Amazonian natives to relate in appropriate ways to their natural surroundings, applying rules of conduct that have sustained life without disturbing nature. The ability of the Indians to take advantage of the possibilities at their disposal is demonstrated through their diet based on protein-poor manioc. One could expect that the natives acquired diseases provoked by improper protein consumption. This does not happen, however, and one finds among the Indians an example of vigorous physical strength (Ribeiro, 1987).

Serious soil deficiencies, on the other hand, have been overcome by elaborate systems of agriculture and intensive soil management. It has been demonstrated by Hecht and Posey (1990) that the Kayapó agricultural system, for example, is superior to modern agricultural methods employed in Amazonia, characterized by pasture and short-cycle crops "which are notorious for their lack of sustainability and low rates of return" (Hecht and Posey, 1990). The Kayapó system does not need purchased inputs and is naturally much richer. The comparison Hecht and Posey (1990) make of Kayapó, colonist, and livestock production patterns of land use in eastern Amazonia reveals that Kayapó yields per hectare, over five years, are more than 180% higher than the yields of the colonist system and over 170 times that of livestock (Hecht and Posey, 1990). The authors also note that in terms of protein yields from vegetable sources over five years, Kayapó figures are roughly double those of colonist agriculture and more than ten times the protein production from livestock, and, "in ten years...1 hectare of pasture has produced less than a ton of meat, and slightly more than 100 kilos of protein, or roughly 5% of the protein generated by the Kayapó system." The conclusion is clear: without damaging the resource base, which modern systems noticeably do, the Kayapó produce much more carbohydrate and protein per hectare than any of the alternatives existing nearby. The irony of the situation is exposed by Hecht and Posey (1990): "Hundreds of millions of dollars have been funnelled into surveys and experiments which have not made the colonists' agriculture more stable, or livestock more productive," yet sitting right on their doorstep is a system whose efficiency they could only dream about. It seems obvious, therefore, that land uses by Amazonian indigenous peoples must mirror in some way or other an endeavor we might call ethnoagronomics. As Hecht and Posey note, "Researchers should also recognize that there is a complex intellectual system that underlies the native management of soil resources, the ensemble of which is 'ethnopedology.'"

It is not surprising to discover with ethnoscience that the kind of itinerant agriculture undertaken by the Indians does not constitute a primitive and incipient method; it is, on the contrary, a specialized technique conceived as a response to specific conditions of climate and soil encountered in the rain forest (Meggers, 1977). Crop diversification, as found in Kayapó territory, equally represents a rational form of land use. The Kayapó have also developed the creation of forest "islands" in tropical savannas, which they term *apêtê*, to modify the ecosystem, thus increasing biodiversity (Museu Goeldi, 1987). This notable ecological engineering is accompanied, for instance, by precise knowledge of insect behavior. A case in point is that the Kayapó deliberately place nests of ants that repel by physical and chemical means leaf-cutting ants in infested gardens and fruit trees (Overal and Posey, 1990).

Indigenous classifications are not aimless. Much to the contrary, they are not only systematic and based on theoretical knowledge, but they are also comparable, from a formal point of view, to those used in modern taxonomy (Lévi-Strauss quoted by Ribeiro, 1987). Using again the example of the Kayapó, we see through Posey's (1992b) description how the Indians classify their natural resources within various ecosystems: "Each ecosystem is perceived by the Indians to exist with a specific association of plants and animals. Having a profound knowledge of animal behaviour, the Kayapó know which plants attract each animal. On the other hand, they associate several species of plants with varieties of soils. Consequently, each ecosystem is a harmonious union of interactions between plants, animals, types of soil and the Kayapó themselves."

Improving soil fertility and productivity is one of the consequences of such a form of classifying ecosystems. When one remembers that modern agricultural practices in Amazonia have exhausted soil fertility and caused serious ecological problems (see Uhl, 1992), the superior ability of the Indians to deal with their environment must be acknowledged. In the case of the Cinta-Larga, anthropologist Carmen Junqueira (1984) has found that all their productive activities obey complex cultural rules which determine the organization of work teams to the different modes of distribution of produce. This complex system of rules and institutions is a counterpoint to technological simplicity and constitutes the pillars of the Indian communal organization (Junqueira, 1984). This same elaborate knowledge is what explains the natives' ability to limit population size, the abhorrent practice of infanticide observed in some groups notwithstanding. Plants such as *Curarea tecunarum* (Ribeiro, 1987) are used by the Deni as a contraceptive, while abstention from sexual activity over long periods after delivery is found among such tribes as the Xamakôko and the Taulipáng (Baldus, 1970) as a means for reducing natality.

The way the Indians understand nature places man as part of a complex network of interactions, including society and the entire universe. This is demonstrated by Reichel-Dolmatoff (1976) in analyzing the meaning of animal behavior to the Indians. What he indicates is that animal behavior represents a model for what is possible, for what can be done for successful adaptation to the environment. "Animals, then, are metaphors for survival. By analyzing animal behaviour the Indians try to discover an order in the physical world, an order to which *human* activities can then be adjusted" (Reichel-Dolmatoff, 1976; his emphasis). The importance of animals for the natives of Amazonia has deep foundations. As food resources, game and fish, together with wild fruits, are viewed in terms of satisfying protein needs. The approach to this evaluation is done with the help of shamans. The Indians equate environmental degradation not with soil exhaustion but with the eventual depletion of game and increased walking times for obtaining food (Reichel-Dolmatoff, 1976). In terms of shamanistic practices, upsetting the ecological balance, by overhunting for instance, is what explains disease. Illness for a Tukano corresponds to a person's interfering with a certain aspect of the ecological order. Incidentally, Reichel-Dolmatoff (1976) remarks that the Tukano as well as several other Amazonian tribes "believe that the entire universe is steadily deteriorating," a clear indication of the Indians' sense of entropy. This tendency can be counterbalanced, according to the Tukano, by a continuous cycle of ritual creation and re-establishment of order and purpose. This is done in ceremonial occasions when the universe and its components are "renewed," and links with past and future generations are reaffirmed.

Ribeiro (1987), referring to the Desâna, explains that despite more than 300 years of contact with the national society, and the corresponding loss of cultural goods, symbols, and values, they continue to treat subsistence by means of wise adaptation to an ecosystem they profoundly comprehend. These and other Indian societies practice austerity in consumption as part of the interaction they perceive between the material sphere and the spiritual world. A close relationship with the principles, cycles, and limits of nature indicates how environmental stress is avoided. In this perspective, nature is not disturbed and the provision of a continuous flow of enough resources for the individuals' well-being is guaranteed. Such a complex system of ecological engineering corresponds to planning life in the limits of the possible, involving present and future generations in the process. This attitude, in turn, amounts to a negation of the non-satiation principle postulated by economists as a normal trait of human character (cf. Faber et al., 1994a,b). It amounts likewise to a holistic way of understanding the world, in acute contrast to the perception of modern man and science. It is interesting to observe that austerity for the Indians does not lead to penury or indigence (poverty is out of the question because it is not applicable as a sociological category to the analysis of Indian society); just the opposite, for an abundance of staples is usually found in the Indian villages. Incidentally, Baldus (1970) reports how lavishly he was received at the Tapirapé village when he arrived for the first time: "Just to give an idea of the food variety of the Tapirapé in a determined period of the year, I want to list the dishes they offered me when, in June 1935, I arrived for the first time at Tampiitaua [23 different dishes are then listed]. Unwilling to offend anyone, I ate in the same afternoon, in all the village's houses, great quantities of each of these delicious dishes."

In 1947 when Baldus returned to the village, this abundance had disappeared, in large measure because of the cultural shock brought about by contact with white people, whose effects modified the formerly unlimited Tapirapé hospitality (Baldus, 1970).

Some Final Thoughts

Following the advice of Seeger (1980) that we should avoid both the evolutionists' ethnocentrism and the romantic view of the noble savage, the Amerindians' lifestyle seems to be a concrete description of how to live sustainably. Certainly it is an extreme situation of compliance with the rules of a sustainable life and a very difficult one to be adopted by modern people. However, the other extreme, epitomized in a recent statement in *The Economist* (November 20, 1993, p. 6), that "to join the rich world means to acquire the ability to grow indefinitely," cannot be seriously considered as a goal to be reached. Georgescu-Roegen (1971) has already demonstrated that no elaborate argument is needed for one "to see that the maximum of life quantity requires the minimum rate of natural resources depletion." To grow forever cannot thus be a global objective to be attained simultaneously and healthily by all countries. The question then is how to imagine a kind of development within the context of the Indian paradigm, of developing within the limits of the possible. As already shown, the Indians study animal behavior precisely as a model for what is possible. Possibilities mean physical constraints, but they also mean the acknowledgment of the second law of thermodynamics, which is an actual limitation even beyond unlimited supplies of resources. The prevailing notion of development associates the pace of natural resource utilization with progress: the higher the pace, the quicker progress takes place (see Tiezzi, 1988). But our way of life, of consumption, also determines the speed of the entropic process, the velocity with which available energy is dissipated. Indian behavior clearly softens the tendency of dissipation. The natives of Amazonia apply naturally, instinctively, the principles of ecology. These same principles could be at the root of the design "of an economic system that can essentially last forever" (Brown, 1991)—last, not grow, forever.

How do the Indians define the forms of their social life? They do not permit an idea like growth to be the center of their preoccupation. "Development" is a purely Western concept (Esteva, 1992) that robs peoples with different cultural frameworks of the opportunity to design their own social objectives. Sustainable development, on the other hand, can occur only if "productive capacity [can be preserved] for the infinite future" (Solow, 1992), that is, if future generations are assured a standard of living not inferior to the present one. Do the Indians preserve productive capacity? Of course they do. They have done so for millenia, not centuries, as the discovery of the "paramount chiefdoms" in Amazonia has proven (Roosevelt, 1991). But the economic performance of the Indians has nothing to do with the Western idea of development. In the case of the Amazon, this conclusion contains an absurdity, for all methods of exploitation of the rain forest that have succeeded the model adopted by the Indians have been revealed as unsustainable. For example, Uhl (1992) notes that "for each cubic meter of wood taken from the forest [with so-called modern methods of production] almost two cubic meters are destroyed." This can be explained by the inconsistent configuration of markets and policies that leaves fundamental resources of life outside the marketplace "unowned, unpriced and unaccounted for—and more often than not [the market] subsidizes [the] excessive use and destruction [of resources] despite growing scarcities and rising social costs" (Panayotou, 1991). To sum up, in Gérald Berthoud's (1992) words, "With money as a supreme value, life counts less." Or, as Gustavo Esteva (1992) says, "Establishing economic values requires the disvaluing of all other forms of social existence." The study of the Indians' lifestyles shows how different the whole picture becomes when life is the supreme value. In this landscape, the emphasis that mainstream economics puts on economic growth before everything else, including distribution, cannot be held. One may look with scorn at a primitive way of life like the

Indians' and consider it simply unacceptable, or a utopia in the modern world. Nevertheless, nothing in nature or society demonstrates that a law of transformation establishes that any given society is in a process of evolution toward "ever more perfect forms" (Esteva, 1992). Or, in Georgescu-Roegen's (1971) view, "no social scientist can possibly predict through what kinds of social organization mankind will pass in its future."

This leads us to the discussion about the need for a paradigm shift away from the dominant model of natural resource (matter and energy) exploitation toward a system of resource use within the earth's carrying capacity and of compliance with the principles of ecology. No doubt, the Amazonian Indians' paradigm offers a proven alternative. This is convincingly illustrated by the example of the Kayapó, on whom ethnobiological research has been conducted since 1977 at the Goeldi Museum in Belém. This research reveals that the Kayapó's "traditional knowledge offers some of the most viable and promising options for sustained resource use in the tropics" (Posey, 1992a). The commitment of the Indian model to the well-being of future generations is another point to be underlined, in accordance with the accepted notion of sustainability (cf. Taylor, 1989). It is also relevant to remember that the Indian paradigm contains an attribute of appreciation for the practical wisdom (Aristotle's *phronesis*—see Faber et al., 1994b) that is meaningful for the solution of environmental problems and the promotion of conservation. It is well known that the market is not reliable in terms of conservation of natural systems. Nothing in its structure induces real sustainability. But not only is sustainability a requirement of the new concept of development, it is also a general prerequisite of life. Goodland (1990) remembers that a voluntary return to sustainability is unavoidable "before global selection does it for us at a much lesser steady-state value."

Particularly for those who live in the Americas, it is extremely important to work with the Amerindian paradigm in mind. It offers an alternative of living sustainably, not to be adopted literally but to be grasped, scrutinized, and understood. It is the opinion of botanists and zoologists doing research with the Indians that the complex relations that ostensibly "primitive" cultures have developed with their surroundings will assume a growing significance in the process of devising policies for the preservation of threatened ecological regions such as the Amazon Basin (Ribeiro, 1987). The emerging body of ethnobiological information in Amazonia shows that ecological sustainability can be attained with the help of indigenous knowledge. Systems of resource management conceived on the basis of such knowledge can promote sustainability "and may generate levels of income that exceed the regional average" (Hecht and Posey, 1990). Thus, there is ample room for paying the greatest attention to the details and intricacies of the Amazon Indians' lifestyle. Theirs is an admirable pattern of co-existence with nature or, more precisely, of knowing how to live within the limits fixed as a challenge by nature.

References

Baldus, H. (1970) *Tapirapé: Tribo Tupí no Brasil Central.* Editora Nacional/Editora da Universidade de São Paulo, Brazil.

Bateson, G. (1972) *Steps to an Ecology of Mind.* Ballantine, New York, NY.

Berthoud, G. (1992) Market. In: *The Dictionary of Development: A Guide to Knowledge as Power.* W. Sachs, Ed. Zed Books, London, U.K., pp. 70–87.

Brown, L. (1991) Is Economic Growth Sustainable? In: *Proceedings of the World Bank Annual Conference on Development Economics, 1991* (roundtable discussion). Supplement to *The World Bank Economic Review* and *The World Bank Research Observer,* pp. 353–355.

Cardim, F. (1939) *Tratados da Terra e da Gente do Brasil.* Editora Nacional, São Paulo, Brazil (originally written in the 16th century).

Cavalcanti, C. (1992) The path to sustainability: austerity of life and renunciation of development, paper presented at the Second Biennial Meeting of the International Society for Ecological Economics, Stockholm, Sweden.

CCPY (Comissão pela Criação do Parque Yanomami) et al. (1990) Yanomami: A Todos os Povos da Terra. Ação pela Cidadania/OAB, São Paulo, Brazil, July.

Chernela, J.M. (1989) Os cultivares da mandioca na área de Uaupés (Tukano). In: *Suma Etnológica Brasileira,* Vol. 1: Etnobiologia. B.G. Ribeiro, Ed. Vozes/FINEP, Rio de Janeiro, Brazil, pp. 151–158.

Cortesão, J. (1943) *A Carta de Pero Vaz de Caminha.* Edições Livros de Portugal, Rio de Janeiro, Brazil.

Daly, H. (1980) Introduction to the steady-state economy. In: *Economics, Ecology, Ethics: Essays Toward a Steady-State Economy.* H. Daly, Ed. Freeman, San Francisco, CA, pp. 1–31.

Esteva, G. (1992) Development. In: *The Dictionary of Development: A Guide to Knowledge as Power.* W. Sachs, Ed. Zed Books, London, U.K., pp. 6–25.

Faber, M., Jöst, F., and Proops, J.L.R. (1994a) Limits and Perspectives of the Concept of Sustainable Development. Discussion Paper 204. University of Heidelberg, Heidelberg, Germany.

Faber, M., Manstetten, R., and Proops, J.L.R. (1994b) Knowledge, Will and the Environment. Discussion Paper 205. University of Heidelberg, Heidelberg, Germany.

Gandavo, P.M. (1924) Tratado da Terra do Brasil. Edição do Annuario do Brasil (originally written in 1570 or earlier).

Georgescu-Roegen, N. (1971) *The Entropy Law and the Economic Process.* Harvard University Press, Cambridge, MA.

Goodland, R. (1990) *Race to Save the Tropics: Ecology and Economics for a Sustainable Future.* Island Press, Washington, D.C.

Gronemeyer, M. (1992) Helping. In: *The Dictionary of Development: A Guide to Knowledge as Power.* W. Sachs, Ed. Zed Books, London, U.K.

Hecht, S. and Posey, D. (1990) Indigenous soil management in the Latin American tropics: some implications for the Amazon Basin. In: *Ethnobiology, Implications and Applications,* Proceedings of the First International Congress of Ethnobiology, Vol. 2. D.A. Posey and W.L. Overal, Eds. Museu Goeldi, Belém, Brazil, pp. 27–49.

Junqueira, C. (1984) Sociedade e cultura: os Cinta-Larga e o exercíco de poder do estado. *Ciência e Cultura,* 36: 1284–1292.

Lewis, M.W. (1992) *Green Delusions: An Environmentalist Critique of Radical Environmentalism.* Duke University Press, Durham, NC.

Meggers, B. (1977) *Amazônia, a Ilusão de um Paraíso.* Civilização Brasiliera, Rio de Janeiro, Brazil.

Mindlin, B. (1994) O aprendiz de origens e novidades. *Estudios Avançados 20* (Universidade de São Paulo), 8: 233–253.

Museu Goeldi (1987) *A Ciência dos Mebengôkre, Alternativas Contra a Destruição.* Museu Paraense Emílio Goeldi, Belém, Brazil.

Overal, W.L. and Posey, D.A. (1990) Uso de formigas *Azteca* spp. para controle biológico de pragas agrícolas entre os Índios Kayapó. In: *Ethnobiology, Implications and Applications,* Proceedings of the First International Congress of Ethnobiology, Vol. 2. D.A. Posey and W.L. Overal, Eds. Museu Goeldi, Belém, Brazil, pp. 219–225.

Panayotou, T. (1991) Is economic growth sustainable? In: *Proceedings of the World Bank Annual Conference on Development Economics, 1991* (roundtable discussion). Supplement to *The World Bank Economic Review* and *The World Bank Research Observer,* pp. 353–355.

Posey, D.A. (1990) The application of ethnobiology in the conservation of dwindling natural resources: lost knowledge or options for the survival of the planet. In: *Ethnobiology, Implications and Applications,* Proceedings of the First International Congress of Ethnobiology, Vol. 2. D.A. Posey and W.L. Overal, Eds. Museu Goeldi, Belém, Brazil, pp. 47–59.

Posey, D.A. (1992a) Introduction to the relevance of indigenous knowledge. In: *Kayapó Science: Alternatives to Destruction.* A. Engrácia de Oliveira and D. Hamú, Eds. Museu Goeldi, Belém, Brazil, pp. 15–18.

Posey, D.A. (1992b) Kayapó science: alternatives to destruction. In: *Kayapó Science: Alternatives to Destruction.* A. Engrácia de Oliveira and D. Hamú, Eds. Museu Goeldi, Belém, Brazil, pp. 19–43.

Reichel-Dolmatoff, G. (1976) Cosmology as ecological analysis: a view from the rain forest. *Man,* 2: 307–318.

Reichel-Dolmatoff, G. (1990) A view from the headwaters: a Colombian anthropologist looks at the Amazon and beyond. In: *Ethnobiology, Implications and Applications,* Proceedings of the First International Congress of Ethnobiology, Vol. 2. D.A. Posey and W.L. Overal, Eds. Museu Goeldi, Belém, Brazil, pp. 9–18.

Ribeiro, B.G. (1987) *O Índio na Cultura Brasileira.* UNIBRADE/UNESCO, Rio de Janeiro, Brazil.

Roosevelt, A.C. (1991) Determinismo ecológico na interpreteção do desenvolvimento social indígena da Amazônia. In: *Origens, Adaptações e Diversidade Biológia do Homem Nativo da Amazônia.* Museu Goeldi, Belém, Brazil, pp. 103–142.

Rostow, W.W. (1960) *The Stages of Economic Growth.* Harvard University Press, Cambridge, MA.

Sachs, W. (1992) One world. In: *The Dictionary of Development: A Guide to Knowledge as Power.* W. Sachs, Ed. Zed Books, London, U.K., pp. 102–115.

Seeger, A. (1980) *Os Índios e Nós. Estudios sobre Sociedades Tribais Brasileiras.* Editoria Campus, Rio de Janeiro, Brazil.

Simon, J. (1987) Now (I think) I understand the ecologists better. *The Futurist,* Sept./ Oct.: 18–19.

Solow, R. (1992) An Almost Practical Step Toward Sustainability. Invited Lecture on the Occasion of the 40th Anniversary of Resources for the Future, Washington, D.C., October 8.

Sousa, G.S. de (1971) *Tratado Descritivo do Brasil em 1587.* Companhia Editoria Nacional/Editoria da Universidade de São Paulo, Brazil (originally written in 1587).

Taylor, P.W. (1989) *Respect for Nature: A Theory of Environmental Ethics.* Princeton University Press, Princeton, NJ.

Tiezzi, E. (1988) *Tempos Históricos, Tempos Biológicos: A Terra ou a Morte: Problemas da "Nova Ecologia"* (translated from Italian by F. Ferreira and L.E. Brandão). Nobel, São Paulo, Brazil.

Uhl, C. (1992) O desafio da exploração sustentada. *Ciência Hoje,* 14: 52–59.

Viveiros de Castro, E. (1986) *Araweté: os Deuses Canibais.* Zahar, Rio de Janeiro, Brazil.

Viveiros de Castro, E. (1992) *Araweté: o Povo do Ipixuna.* Centro de Documentação e Informação, São Paulo, Brazil.

Ecologically Sustainable Institutions

3

Aseem Prakash
Department of Strategic Management and Public Policy,
School of Business and Public Management,
The George Washington University, Washington, D.C.

Anil K. Gupta
Indian Institute of Management, Ahmedabad, India

Abstract

In this chapter, we explore whether self-interested humans can create institutions to equitably share resources with future generations as well with other species. Institutions are rules that permit, prescribe, or prohibit certain actions. Two categories of institutions are discussed: (1) social institutions governing interaction among humans and (2) ecological institutions governing the interactions of humans with other species. Ecologically sustainable institutions may be conceptualized using the Rawlsian *veil of ignorance* (Rawls, 1971). Some examples from India that implicitly use Rawls' principle to devise such institutions are described. Emphasis is on the key role of cultural and religious institutions in enabling individuals to reinterpret their notion of self-interest and to create incentives to harmonize their self-interest with that of future generations and other species.

Introduction

This chapter explores whether rational and self-interested humans can create rules to equitably share resources with future generations as well with other

1-57444-077-2/97/$0.00+$.50
© 1997 by CRC Press LLC

species. Humans are social beings as well as ecological beings. The social aspect of human existence is reflected, *inter alia,* in the need for collective action. Collective action refers to those situations where certain objectives cannot be achieved or can only inefficiently be achieved by individual effort alone and the cooperation of other people is required (Wade, 1987). To facilitate collective action, humans construct social "institutions"; that is, they create norms, rules, and governance structures so that group members may accomplish a common goal (Ostrom, 1990; North, 1990). Institutions may also be used as instruments of power to establish domination over others (Libecap, 1989; J. Knight, 1992). Institutional design may therefore reflect considerations of both efficiency and power. In this chapter, two categories of institutions are discussed: (1) social institutions governing the interaction among humans and (2) ecological institutions governing the interaction of humans with other species. Often the same institution may influence the social relationships to serve as a social institution and influence the ecological relationships to serve as an ecological institution.

We relate our notion of institutions to the issue of ecological sustainability. Ecological sustainability has two distinct dimensions: (1) a "fair" division of resources between current and future generations (World Commission on Environment and Development, 1987) and (2) a "fair" division of resources among humans and other species. In this chapter, we explore whether humans can devise social and ecological institutions to ensure a "fair" division of resources with future generations as well with other species, especially if what constitutes a fair distribution is contested. We argue that dominant schools of economics, neoclassical economics and new-institutional economics, focus on only a subset of "social" aspects of institutions and totally neglect the "ecological" aspect. Hence, the ecological institutions in these branches of economics are anthropocentric and the social institutions appear to be biased against future generations. Public policies based on contemporary economics may therefore not serve the objective of ecological sustainability.

Humanity needs to reconceptualize social and ecological institutions to alter its dominance over future generations and other species. This is challenging, as there are different notions on the gravity of the ecological crisis (some question even its very existence) and the strategies to tackle it (e.g., J.L. Simon, 1981; Nordhaus, 1994). The environmental policy discipline, which deals with the issues of ecological sustainability, lacks sufficient theoretical perspective and overemphasizes descriptive and normative concerns (Francis, 1990; Fairfax and Ingram, 1990).

Ecological economics appears to be an exception to this trend in that it offers a theoretical perspective on how scholars may analyze environmental problems and conceptualize ecologically sustainable institutions (see, e.g., Daly, 1992a,b). As an interdisciplinary field, ecological economics combines insights from ecology and economics to explicitly focus on issues of ecological sustainability. The institutional focus of ecological economists, however, remains relatively underdeveloped. In particular, ecological economists have insufficiently explored how institutions may be constructed to deal with the multiple objectives of efficiency, equity, and sustainability.

In the last section of this chapter, we discuss how ecologically sustainable institutions may be conceptualized using the Rawls' concept of a *veil of ignorance* (Rawls, 1971). This concept offers a theoretical perspective on how rational self-interested individuals can equitably divide resources among themselves, future generations, and other species. We describe some examples from India that implicitly use Rawls' principle to devise such institutions. We emphasize the key role of cultural and religious institutions in enabling individuals to reinterpret their notion of self-interest and creating incentives to harmonize their self and collective interests. We emphasize that Rawls' principle is only a starting point in conceptualizing and designing social and ecological institutions that create incentives for the pursuit of ecological sustainability. Rawls' principle will need to be operationalized by individuals at different levels of resource aggregation such as global, regional, national, provincial, village, household, etc. Further, such operationalization will need to be culture specific.

Collective Action and Social Institutions

Humans devise rules, norms, and governance structures to regulate their social relationships. Such rule-ordered relationships are known as social institutions (Young, 1982; Ostrom, 1986, 1990; Keohane, 1989; Keohane et al., 1993). Social institutions have been described as "collective action in control of individual action" (Common [1934] 1959, pp. 69–70), or sets of rules that order repetitive and interdependent activities among individuals and highlight what actions are required, prohibited, or permitted (Ostrom, 1986). Institutions are different from organizations. Institutions are rules, whereas organizations are physical actors (North, 1990). For example, the courts of law are organizations, whereas the laws themselves are institutions. Institutions may be formal or informal. What is important is that they constrain human action by changing the incentive structure facing the various actors.

Social institutions facilitate collective action. Collective action refers to those situations where an action is undertaken by more than one individual to achieve a common goal and this goal cannot be achieved efficiently by individual action alone (Wade, 1987). This typifies everyday issues regarding the provision and consumption of many of goods and services. Collective action is difficult to organize in that individual interest and the collective interests may not be in

Table 3.1   The Prisoner's Dilemma Game: Payoff Matrix

	Prisoner A	
Prisoner B	*Not confess (NC)*	*Confess (C)*
Not confess (NC)	(NC,NC) (2,2)	(NC,C) (4,1)
Confess (C)	(C,NC) (1,4)	(C,C) (3,3)

harmony. Such situations are called collective action dilemmas because the pursuit of individual interest may not lead to socially efficient outcomes. Efficiency or "Pareto efficiency," refers to those situations where it is not possible to increase the welfare of any individual without making someone else worse off.

The key aspect of collective action situations is the interdependence among actors. Game theory therefore becomes a useful tool in understanding the theoretical basis of such interdependence and how institutions may be constructed so that self-interested but interdependent individuals may attain socially efficient outcomes (Ostrom, 1990). Game theorists differentiate between two categories of collective action dilemmas: the coordination dilemmas (e.g., the Assurance Game) and the cooperation dilemmas (e.g., the Prisoner's Dilemma Game) (Stein, 1982). The two games are illustrated in Tables 3.1 and 3.2.

In the Prisoner's Dilemma Game (Table 3.1), two prisoners are assumed to be placed in separate cells where they cannot communicate. The sheriff gives both prisoners identical options: (1) if both prisoners confess, he will ask the court for a three-year prison sentence for both of them; (2) if one confesses and the other does not, he will ask for a one-year sentence for the squealer and four years for the other; (3) if both do not confess, he will ask for two years for both of them. Consider the logic adopted by prisoner A: if B confesses, then it is rational for him to confess (three years vs. four years); if B does not confess, it

Table 3.2   The Assurance Game: Payoff Matrix

	Prisoner A	
Prisoner B	*Not confess (NC)*	*Confess (C)*
Not confess (NC)	(NC,NC) (2,2)	(NC,C) (4,3)
Confess (C)	(C,NC) (3,4)	(C,C) (2,2)

is still better to confess (one year vs. two years). Thus, for prisoner A, the optimal strategy is to confess. Both A and B follow the same logic and confess (three years for both). However, both would have been better off not confessing (two years each).

In the Assurance Game (Table 3.2), the payoff structure is different from the Prisoner's Dilemma Game. The two prisoners are given the following options: (1) if both confess or if both do not confess, both get two years each; (2) if one confesses and the other does not, the squealer gets three years and the other gets four years. Let us consider the logic adopted by A. If B does not confess, the best strategy is not to confess (two years vs. three years); if B confesses, the best strategy is to confess (four years vs. two years). Thus, both prisoners, acting in a rational self-interested manner, need to either confess or not confess. If they do not adopt the same strategy, then both may serve longer sentences.

The two games represent different categories of collective action dilemmas. The Assurance Game represents a "coordination dilemma" in that the self-interests of the actors are consistent; "unilateral confession" (three years) is Pareto-inferior to "mutual non-confession" (two years) as well as to "mutual confession" (two years). Hence, there are two equilibria: mutual confession and mutual non-confession. The challenge is to choose from the two equilibria, and rules are crafted to facilitate this choice.

The Prisoner's Dilemma, in contrast, represents a "cooperation dilemma" in that the interests of the actors do not converge. Unilateral confession (one year) is Pareto-superior to mutual non-confession (two years). However the *raison d'etre* for cooperation exists as mutual confession (three years) is Pareto-inferior to mutual non-confession (two years). For mutual non-confession to succeed, both actors need to constrain their opportunistic tendencies to unilaterally confess. To monitor and sanction such opportunism, actors create rules. The term "institutions" is used to describe enforced rules that facilitate the identification of an equilibrium as in coordination dilemmas or that constrain opportunism in cooperation dilemmas.

Other types of games can be constructed by varying the payoff matrix. The foregoing examples are one-shot games. We can expect different outcomes for repeated games in which actors have to deal with each other repeatedly over a longer period of time (Ostrom, 1990). Further, the above games do not allow for communication among the actors. If we allow for communication between the prisoners, the outcomes can potentially change (Ostrom et al., 1994). The critical insight from game theory is that the outcomes in situations where self-interested individuals are interdependent depend critically on the payoff structure, the number of actors, the ability to communicate, whether the actors will confront each other in future, and so on. Most of these factors can be influenced by consciously crafted institutions. Thus institutions have the potential to chan-

Table 3.3   Categories of Goods and Services

	Rivalrous	*Non-rivalrous*
Excludable	Private goods	Toll or club goods
Non-excludable	Common-pool resources	Public goods

Source: Ostrom et al., 1994.

nel human self-interest into socially efficient as well as socially inefficient outcomes.

The attributes of goods and services also influence collective action dilemmas. Political economists often classify goods and services along two dimensions: rivalry and excludability in consumption. Consumption is considered to be rival if A's consumption of a particular unit of the good precludes B's consumption. Excludability implies that it is technologically possible for A to prevent other actors from consuming that particular unit of good or service and this can be done at a low cost to A. Excludability depends on factors such as the physical nature of the good, technology to ensure excludability, effectiveness of the property rights regime, etc. Based on these two attributes, goods and services may be classified into four stylized categories: public goods, private goods, common-pool resources, and club or toll goods (Ostrom et al., 1994). Public goods, club goods, and common-pool resources are also called collective goods, signifying that collective action is key to their provision and/or consumption. These are presented in Table 3.3.

Non-excludability and rivalry create different types of collective action dilemmas. Non-excludability creates incentives for "free riding" (Olson, 1965) and consequently leads to the underprovision of the good (Ostrom et al., 1994). Since no actor can be excluded from partaking the benefits, rational actors do not invest resources to provide the good in the first place. Markets elicit contributions toward provision through the threat of exclusion. Since markets may not be effective for provisioning non-excludable goods and services, some alternative institutional mechanism is required.

Rivalry in consumption creates incentives for the overuse of resources and rent dissipation (Ostrom et al., 1994). If a good or service is rivalrous and excludable (as in a private good), then overconsumption is checked through a rise in price, which reduces the excess demand. However, if the rivalrous good is non-excludable (as in a common-pool resource), then markets do not operate and scarcity of the resource does not translate into a reduction in demand. Non-market institutions are therefore necessary to ensure that the resource is used in a sustainable manner.

Natural resources such as fisheries, water resources, and grazing lands have the characteristics of common-pool resources in that their consumption is rival but exclusion is difficult or costly to enforce. If the consumption of such resources is not regulated, they become "open access" systems and lead to the so-called "tragedy of commons" (Hardin, 1968). This eventually harms the people who depend on open-access resources for their livelihood. Social institutions are therefore constructed to restrain the overexploitation of such resources. This restraint is not altruistic. It is governed by self-interest in that the resources are now available for use over a longer period of time.

We have discussed various kinds of collective action dilemmas and how institutions can potentially mitigate them. We now explore the various views on the emergence of social institutions. Social institutions may be spontaneous, negotiated, or imposed (Young, 1982). Spontaneous institutions emerge without any conscious planning by the actors. Though such institutions result from the actions of many actors, they are not informed by any deliberate design (Hayek, 1973). The expectations of participants converge around certain "focal points" even though there is no explicit communication on what such points may be (Schelling, 1960). Such focal points embody the mutual expectations of actors and enable them to coordinate their actions.

Negotiated institutions, on the other hand, have a contractarian character (Bates, 1988). They are conscious human artifacts in that they emerge as a result of deliberation and bargaining. Imposed institutions are dictated by the stronger actors, often to the detriment of the weaker actors. Imposed institutions may also be viewed as a subcategory of contractarian institutions in which the various actors have asymmetrical bargaining power.

Of course, the categorization of institutions as spontaneous, contractarian, and imposed is stylized. These categories may best be viewed as lying along a continuum with the contractarian institutions in the middle and the imposed and spontaneous institutions on the two ends. However, it is important to identify whether an institution is predominantly spontaneous contractarian or imposed in order to understand the power relationships and whose interests an institution serves.

In the contractarian perspective, institutions are viewed as human artifacts created to reduce transaction costs (Coase, 1937; North, 1990; Williamson, 1975, 1985). Transaction costs pertain to costs of searching for information, processing information, negotiating contracts, monitoring contracts, and sanctioning if contracts are violated (Eggertsson, 1990). However, this exclusive focus on efficiency may lead us to ignore the power relationships. There are many ways in which the members of a group can cooperate to enhance group efficiency. The choice of a particular route often reflects the power relationships in social institutions. Power does not imply absolute control by one actor over another. Power is a relational concept (Hart, 1976). If A exercises power over B, then A can alter the preferences and/or strategies of actor B. Of course, both A and B often exert power over each other at the same time; there is seldom a one-sided exercise of power (Prakash and Ostrom, 1996). Even in imposed institutions, the weaker party exercises some power over the stronger party. In a tyrannical dictatorship, the subjects have some power over the dictator. Dictators often feel insecure about the loyalty of their subjects and invest resources in propaganda, surveillance, and other mechanisms to maintain this loyalty.

A chronic imbalance of power may lead to socially inefficient outcomes. The more powerful actors may sustain institutions that safeguard their power at the cost of the rest of the participants. North (1990) uses a similar logic to explain the persistence of regimes that serve the ruler but prey on citizens and undermine the overall welfare of the society. Thus, both the politics (distribution of costs and benefits) and economics (efficiency through cooperation) are important to understand the emergence and sustenance of social institutions.

Ecological Institutions

Ecological institutions are rules, norms, and governance structures that regulate the interactions of humans with other species. Conceptually, ecological institutions, like social institutions, may be classified as spontaneous, contractarian, and imposed. However, ecological institutions have a *de facto* imposed character in that they are often designed to serve human interests only. Ecological institutions cannot have a contractarian character. For contracting to take place, all the participants must share a common language, a condition absent between humans and other species.

Social institutions may have ecological dimensions. For example, markets as social institutions value resources according to the preferences of people. This has two implications. First, even if such resources are needed by other species, they are unable to articulate their preferences. Second, other species themselves become resources assessed only for their instrumental value. The challenge now is to construct ecological institutions that safeguard the interests of other species even though they may not participate in constructing such institutions.

Economics and Institutions

Embedded in the various schools of economics are specific notions about social and ecological institutions. We focus on the two most influential schools of contemporary economics, neoclassical and new-institutional, as well as on the emerging field of ecological economics.

Economics deals with the allocation of scarce resources (Robbins [1932] 1937). Neoclassical economics, also called price theory, gives primacy to the social institution of the market for allocating resources. The "invisible hand" (Smith [1776] 1937) of the market, if allowed to operate without any restrictions, is supposed to maximize the efficiency of resource use. In such a perspective, there are no collective action dilemmas, humans are perfectly rational with stable and transitive preferences, there is free flow of information at zero transaction costs, and markets work perfectly.

Markets as social institutions facilitate specialization and exchange (De Alessi, 1993). Hayek (1945) underlines the role of the market in communicating information among economic actors on the relative scarcity of resources. The market conveys information to various actors in the form of prices of various resources, and such prices are determined by the preferences of the current participants in the exchange process. It may be argued that market prices incorporate the preferences of future generations through the instrument of the discount rate. There is a substantial literature on the determinants of discount rates and why people prefer the present over the future (e.g., F.H. Knight [1933] 1964; Robinson, 1979; Fisher, 1986). We see the discount rate being determined by the preferences of current participants about their own future. It is seldom determined by the current generation acting to maximize the interest of future generations and, of course, it cannot be determined by the future generations themselves. Clearly, the discount rate, as used currently, is a poor instrument to ensure an equitable use of resources across generations. Further, markets are also inequitable ecological institutions in that prices seldom reflect the preferences of non-human species.

Neoclassicalists focus on only one type of social institution, the market, which is considered to evolve spontaneously as a result of the inherent tendency of people to barter and exchange (Hayek, 1945).* New-institutionalists, by contrast, consider social relationships to be regulated by a variety of institutions, markets being only one of them. New-institutionalists retain the assumptions of stable, transitive, and exogenous preferences, as well as of rational utility maximization. However, unlike neoclassicalists, new-institutionalists view actors as "boundedly rational" due to constraints on neurological as well as transaction costs (Simon, 1957; Ostrom, 1990; Eggertsson, 1990). Complete rationality means that actors have perfect information about current and future contingencies. Bounded rational actors, by contrast, cannot perfectly anticipate current and

---

* Polanyi ([1944] 1957) has argued that markets have not emerged spontaneously and are conscious human artifacts to serve particular ends.

Table 3.4   **Institutions and Economics**

	Neoclassical	*New-institutional*	*Ecological*
Social institutions	Only market	Market, state, self-organized units, etc.	Market, state, self-organized units, etc.
Ecological institutions	Anthropocentric	Anthropocentric	Biocentric

future contingencies. This creates potential for cheating or "opportunism" as Williamson (1975) puts it. To safeguard against potential opportunism, elaborate and restrictive contracts need to be drawn, and this creates transaction costs. Institutional designs reflect the desire to minimize transaction costs (Williamson, 1975, 1985).

Both neoclassicalists and new-institutionalists are anthropocentric in that they focus only on interactions among humans and ignore the implication for other species. They thus ignore the ecological aspects of social institutions. Ecological economists, in contrast, focus on the ecological dimension of social institutions. Ecological economics is viewed as a "...trans-disciplinary field of study that addresses the relationships between the ecosystem and the economic system...it differs from both conventional economics and conventional ecology in terms of both the breadth of its perception of a problem, and the importance it attaches to environment–ecology interactions" (Costanza et al., 1991).

Ecological economists assume a biocentric ecological ordering. Non-human species are assumed to have an existence value and they are not assessed for their instrumental value alone. A symmetric relationship among species is normatively assumed. The challenge for ecological economists is to move from the normative sphere to the positive sphere. Specifically, they need to explicitly lay out how ecologically sustainable institutions may be constructed so that current and future generations, as well as humans and other species, get a "fair" share of earth's resources. A major challenge is to explore whether such institutions are consistent with the notion of a self-interested individual. Rawls' *veil of ignorance* is a useful starting point in this context, and we elaborate on this in the last section of this chapter. A summary of our discussion on the three perspectives of economics is provided in Table 3.4.

Sustainable Institutions

Social institutions often have multiple objectives, such as efficiency, equity, and sustainability (Ostrom et al., 1993). Efficiency or Pareto-efficiency, as defined above, implies that it is impossible to increase the welfare of any actor without decreasing the welfare of any other actor. Equity refers to the allocation of resources among people. Social institutions typically pursue equity through two mechanisms: fiscal equivalence and ability to pay (Ostrom et al., 1993). Fiscal equivalence means that those who get greater benefits are expected to pay more for them. The ability to pay means that better-off individuals pay more than poorer individuals for every unit of benefit they receive. The ability-to-pay principle may also be interpreted as a redistributive strategy in that the better-off members subsidize the other members.

Sustainability refers to allocation of resources across generations or between human and other species. Often, the objectives of efficiency, equity, and sustainability do not harmonize. To begin conceptualizing ecologically sustainable social institutions, we first need to prioritize the objectives of efficiency, equity, and sustainability and establish the trade-offs between them. Since actors may have different preferences for such prioritization, we also need to identify mechanisms through which common ground may be established.

Daly (1992a,b), building on the work of Tinbergen ([1952] 1966), argues that to achieve the three "independent" objectives of efficiency, equity, and sustainability, three "independent" policy instruments are needed. Further, these policy instruments need to operate in different social institutions. Efficiency is to be pursued through the price mechanism operating in the institution of the market. Equity is to be pursued through the instrument of fiscal policy, which is administered through the institution of the state. Sustainability is to be achieved through the instrument of scale. By scale Daly means an upper limit to resource use. Daly does not, however, identify the appropriate social institution to pursue sustainability, nor does he explain who will decide the appropriate scale. Daly's thesis is summarized in Table 3.5.

Some scholars argue that the market is the most appropriate institution for pursuing ecological sustainability. The new resource economists in particular argue that private ownership of resources, coupled with competitive markets, ensures sustainable resource use (Anderson and Leal, 1991). Further, they maintain that even if property rights are imperfectly delineated, the use of markets to

Table 3.5 Goals, Instruments, and Institutions

Goal	Instrument	Institution
Efficiency	Price mechanism	Market
Equity	Fiscal policy	State
Sustainability	Scale	

facilitate bargaining between generators and receivers of externalities will result in an efficient internalization of externalities (Coase, 1960).

We have mentioned that goods and services may be classified along two dimensions: rivalry and excludability. Building on the same classification, environmental sustainability can exhibit the characteristics of a private good, a public good, a common-pool resource, or a toll good. Typically, sustainability is a non-excludable good in that people cannot be prevented from enjoying the benefits of ecological balance. We have also considered that the provision of any non-excludable good faces collective action dilemmas and that the institution of market fails to optimally provide for it. An exclusive reliance on markets for provisioning sustainability seems untenable.

In standard public policy literature, market failures provide a rationale for state intervention (e.g., Weimer and Vining, 1989). State intervention takes various forms, such as an imposition of a tax on the polluter equal to the divergence between the social and private cost (Pigou [1932] 1960), specification of maximum pollution levels through command-and-control policies, etc. The logic for state intervention to correct market failures is flawed. The deficiency of one social institution (the market) does not imply the superiority of some other social institution (the state) (Ostrom and Walker, 1993). The standard public choice critique is that if markets are susceptible to market failures, then state intervention too is susceptible to government failures (Wolf, 1979; Müller, 1989). It is difficult to speculate, *a priori*, which failure is more virulent.

Daly assumes that the three objectives of efficiency, equity, and sustainability are independent of each other. Based on these assumptions, he adopts Tinbergen's ([1952] 1966) argument of "three independent goals and therefore three independent instruments."* The efficiency–equity–sustainability independence is questionable.** The efficiency–equity connection has been debated at length (e.g., see Okun, 1975). Norgaard (1992), on the other hand, does not see any trade-off between efficiency and sustainability. For him, the efficiency is not

---

* For a critique of Daly's thesis on treating sustainability, equity, and efficiency as independent goals, see Prakash and Gupta (1994).

** Also, when Daly talks of the market, he implies a competitive market. Such markets seldom exist in reality. Neoclassicalists have a strong argument for equating such markets with efficiency. However, if firms (a hierarchy) are replacing markets, then why not rely on the state (another hierarchy) to pursue allocative efficiency (Lange, 1938)? As Pollack (1985) has pointed out, a corporate hierarchy *per se* is not superior to a governmental hierarchy on transactional cost considerations. It may be argued that the principal agent problems are controlled in firms as they are regulated by the stock market (Manne, 1967; Jensen and Meckling, 1977; Fama, 1980). Daly's thesis is problematic in that he differentiates a firm-dominated exchange process from a state-dominated one and equates only the former with efficiency.

context independent. For every distribution of assets, there is a corresponding efficient allocation of distribution. However, unlike Daly, Norgaard treats sustainability as an issue of intergenerational equity.

So what is the alternative? Are there non-state and non-market institutions capable of efficient provisioning for environmental sustainability? Self-governing structures constitute such a genre of institutions. The members of a self-governing unit, without recourse to any external agent such as the state or the market, devise, monitor, and enforce rules for resource use so that private and social costs converge, overuse is curbed, and free riding is checked (Ostrom, 1990). Such institutions are particularly useful in the case of communally owned common-pool resources where the livelihood of the actors critically depends on the sustainable use of resources.

Rawls' Veil of Ignorance

There is an extensive literature that interprets environmental sustainability as putting an upper limit on resource use by human beings. Hardin's *carrying capacity* (1968), Kapp's *social minimas* (1970), Dietz and Straaten's *ecologically bounded possibilities* (1992), and Daly's *Steady-State Economics* (1992b) emphasize the same notion. Such concepts, however, do not provide a theory to link human self-interest to environmental sustainability, specifically, how collective action dilemmas in the provisioning of environmental sustainability (nonexcludable and hence susceptible to free riding) may be overcome. A major challenge for ecological economists and other scholars dealing with issues of environmental sustainability is to develop a theoretical perspective in which social and ecological institutions create incentives for self-interested individuals to equitably divide resources among themselves, future generations, and other species. Such an enterprise will face the following challenges: (1) how to develop a shared understanding of what constitutes an equitable distribution and the processes for arriving at such an understanding and (2) how to posit preferences for future generations and for other species regarding resource sharing.

The *veil of ignorance* (Rawls, 1971) concept is a useful starting point. Rawls, like the new-institutionalists, regards humans to be self-interested. He also sees institutions to specify the payoffs corresponding to the various strategies available to actors. Rawls posits a specific kind of institutional setting to create incentives for the stronger actor to equitably divide resources. Rawls' institutional conception is of the following kind. Assume that the stronger actor is asked to divide a particular resource in two parts. Since the weaker actor is assumed to exercise the first choice, the stronger actor divides the resource without any *ex ante* knowledge of which part will accrue to him. The stronger actor is therefore assumed to operate under a *veil of ignorance*. Given this uncertainty, the utility-maximizing strategy for the stronger actor is to equally divide the resource. This is also called the maxi-min strategy in that the stronger actor maximizes the minimum state.

A major advantage of using this principle is that we are not required to posit preferences of future generations and other species. Further, although Rawls assumes that all actors are contemporaries, this construct holds even when actors belong to different generations and different species (Page, 1977; Tacconi and Bennett, 1995). However, we face a conceptual problem in extending this principle from a limited-actor situation to an infinite-actor situation. Rawls assumes the number of actors to be finite and known. If the actors are infinite or their number unknown, then the stronger actor cannot know whether the division of resources is equitable or not. In such cases, the stronger actor needs to assume what he considers to be relevant generations and species for constructing social and ecological institutions.

Rawls' principle does not provide us with a blueprint or an organizational design to construct equitable social and ecological institutions. Rather, it offers a theoretical perspective on how the self-interest of a stronger actor (humans) may be harmonized with that of a weaker actor (future generations and other species).* Depending on the nature of the resource, this principle needs to be operationalized at different levels of human aggregation, such as international, regional, national, local, etc. This requires that people must have the autonomy to frame their own rules. Since individuals will need to operationalize this principle in a culture-specific manner, we see the evolution of institutions based on the *veil of ignorance* as leading to a decentralized system of governance.

Our proposal for designing institutions based on the *veil of ignorance* is practicable. Many people implicitly use this principle in their daily lives while giving it their own cultural interpretation. Gandhi's conception of "trusteeship," the stronger entity managing a common resource on behalf of the larger community, reflects this principle.** The weaker entities are presumed to have trusted the management of such resources to the stronger entity to be used for the common good. We also see the creation of nature reserves and the declaration of some species as "endangered" as examples where the current generation has reinterpreted its self-interest to give preference to the interests of other species.

---

* There is an extensive literature on the "veil of ignorance" as a basis to explore intergenerational equity. See, for example, Weiss (1989, 1990), D'Amato (1990), Gunding (1990), and Tacconi and Bennett (1995).
** See Gandhi (1957), Fischer (1950), and Upadhyaya (1976) on the philosophy of trusteeship.

In India, we find instances where starving people continue with their daily routine of feeding animals over feeding themselves. Hindus and some animist traditions view certain animals to be "sacred." The needs of such animals supersede the needs of humans. Such examples appear to contradict the model of a self-interested individual maximizing his utility. However, we see religion as a key factor in enabling individuals to re-interpret "self-interest." For example, Hindus believe in the reincarnation of the soul. It is believed that the "books of accounts" of every individual based on his or her *karmas* (actions) are balanced across incarnations (8.4 million in the case of Hindus). From this perspective, the stronger creature has a "self-interest" in adopting a benign attitude toward weaker beings to avoid being sanctioned in future incarnations. Many mythological tales and folklores describe how a stronger entity that exploited the weak was condemned to the life of a helpless creature in its future incarnations. Such tales and folklores thus reinforce the desirability of "doing good" in the current incarnation to avoid punishments in future incarnations. Religious and cultural practices provide examples where humans apply the *veil of ignorance* in their everyday routines. Since religion is a set of humanly constructed institutions (at least according to us), one reason for the evolution of such actions, we hypothesize, is concern for the ecological balance.

Tocqueville ([1835] 1945) emphasized that "self-interest rightly understood" forms the basis for a democratic society. We think the *veil of ignorance* principle can help the current generation to make a transition from destructive self-interest to an enlightened self-interest and construct ecologically sustainable institutions.

Acknowledgments

This is a revised version of a paper presented at the 3rd International Conference of the International Society for Ecological Economics in San José, Costa Rica, October 3–8, 1994. We thank Brenda Bushous, Sue Crawford, Ray Elison, Alexander Farell, and Fraser Smith for their comments.

References

Alchian, A. and Demsetz, H. (1972) Production, information costs, and economic organization. *American Economic Review,* 62: 777–795.

Anderson, T. and Leal, D.R. (1991) *Free Market Environmentalism.* Westview Press, Boulder, CO.

Arrow, K.J. (1963) *Social Choice and Individual Values.* Wiley, New York, NY.

Bates, R.H. (1988) Contra contractarianism: some reflections on new institutionalism. *Politics & Society,* 16(2–3): 387–401.

Baumol, W.J. and Oates, W.E. (1988) *The Theory of Environmental Policy,* second edition. Cambridge University Press, Cambridge, U.K.

Cheung, S.N.S. (1970) The structure of a contract and the theory of non-exclusive resource. *Journal of Law and Economics,* 49: 49–70.

Cleveland, C.J. (1991) Natural resource scarcity and economic growth revisited—economic and biophysical perspectives. In: *Ecological Economics: The Science and Management of Sustainability.* R. Costanza, Ed. Columbia University Press, New York, NY, pp. 289–317.

Coase, R.H. (1937) The nature of the firm. *Economica,* 4: 386–405.

Coase, R.H. (1960) The problem of social cost. *Journal of Law and Economics,* 3: 1–40.

Common, J.R. ([1934] 1959) *Institutional Economics: Its Place in Political Economy.* University of Wisconsin Press, Madison, WI.

Costanza, R., Daly, H.E., and Bartholomew, J.A. (1991) Goals, agendas and policy recommendations for ecological economics. In: *Ecological Economics: The Science and Management of Sustainability.* R. Costanza, Ed. Columbia University Press, New York, pp. 1–21.

Daly, H.E. (1990) Ecological economics and sustainable development: ecological physical chemistry. In: Proceedings of an International Workshop, Nov. 8–12, 1990, Siena, Italy, pp. 185–201.

Daly, H.E. (1992a) Allocation, distribution, and scale—towards an economics that is efficient, just, and sustainable. *Ecological Economics,* 6(3): 185–193.

Daly, H.E. (1992b) *Steady-State Economics,* second edition. Earthscan Publications, London, U.K.

D'Amato, A. (1990) Do we have a duty to future generations to preserve the global environment? *American Journal of International Law,* 84(1): 190–198.

De Alessi, L. (1993) How markets alleviate scarcity. In: *Rethinking Institutional Analysis and Development.* V. Ostrom, D. Feeny, and H. Picht, Eds. ICS Press, San Francisco, CA, pp. 340–376.

Dietz, F.J. and Straaten, J.V.D. (1992) Rethinking environmental economics. *Journal of Economic Issues,* 26(1): 27–51.

Eggertsson, T. (1990) *Economic Behavior and Institutions.* Cambridge University Press, Cambridge, U.K.

Fairfax, S.K. and Ingram, H. (1990) No theory, no apology: a brief comment on the state of the art in natural resource policy and the articles herein. *Natural Resources Journal,* 30(2): 259–262.

Fama, E. (1980) Agency problems and the theory of firm. *Journal of Political Economy,* 88: 288–307.

Fischer, L. (1950) *The Life of Mahatma Gandhi.* Harper, New York, NY.

Fisher, I. (1986) *The Theory of Interest Rates.* A.M. Kelly, Fairfield, NJ.

Francis, J.G. (1990) Natural resources, contending theoretical perspectives, and the problem of prescription: an essay. *Natural Resources Journal,* 30(2): 263–282.

Gandhi, M.K. (1957) *An Autobiography: The Story of My Experiments with Truth.* Beacon Press, Boston, MA.

Gunding, L. (1990) Our responsibility to the future generations. *American Journal of International Law,* 84(1): 207–212.

Gupta, A.K. and Prakash, A. (1992) Choosing the Right Mix—Market, State, and Institutions for Environmentally Sustainable Industrial Growth. Working Paper #1066. Indian Institute of Management, Ahmedabad, India.

Gupta, A.K. and Prakash, A. (1993) On Internalization of Externalities. Working Paper #1126. Indian Institute of Management, Ahmedabad, India.

Hardin, G. (1968) The tragedy of commons. *Science,* 162: 1243–1248.

Hart, J.A. (1976) Three approaches to the measurement of power in international relations. *International Organization,* 30: 289–305.

Hayek, F.A. (1945) The use of knowledge in society. *American Economic Review,* 35(4): 519–530.

Hayek, F.A. (1973) *Rules and Order,* Vol. 1: Law, Legislation, and Liberty; quoted in Young, O.R. (1989) *International Cooperation—Building Regimes for Natural Resources and the Environment.* Cornell University Press, Ithaca, NY, pp. 84–85.

Jancar-Webster, B. (1987) *Environment Management in the Soviet Union and Yugoslavia.* Duke University Press, Durham, NC.

Jensen, M.C. and Meckling, W.H. (1977) Theory of the firm: managerial behavior, agency costs, and ownership structure. *Journal of Financial Economics,* 3: 305–360.

Kapp, W.K. (1970) Environmental disruption—general issues and methodological problems. *International Social Sciences Council,* 8(4): 15–32.

Keohane, R.O. (1989) *International Institutions and State Power.* Westview Press, Boulder, CO.

Keohane, R.O., Haas, P.M., and Levy, M. (1993) The effectiveness of international environmental institutions. In: *The Institutions for Earth—Sources of Effective International Environmental Cooperation.* P.M. Haas, R.O. Keohane, and M. Levy, Eds., MIT Press, Cambridge, MA, pp. 3–24.

Knight, F.H. ([1933] 1964) *Risk, Uncertainty, and Profit.* Augustus M. Kelly, New York.

Knight, J. (1992) *Institutions and Social Conflict.* Cambridge University Press, Cambridge, U.K.

Lange, O. (1938) *On Economic Theory of Socialism.* University of Minnesota Press, Minneapolis, MN.

Libecap, Ga. (1989) *Contracting for Property Rights.* Cambridge University Press, Cambridge, U.K.

Manne, H.G. (1967) Mergers and markets for corporate control. *Journal of Political Economy,* 73:110–120.

Manser, R. (1993) *Failed Transition: The Eastern Europe Economy and Environment Since the Fall of Communism.* New Press, New York, NY.

Müller, D.C. (1989) *Public Choice II.* Cambridge University Press, Cambridge, U.K.

Neale, W.C. (1987) Institutions. *Journal of Economic Issues,* 21(3): 1177–1206.

Nordhaus, W.D. (1994) Climate change and economic development. In: *Proceedings of the World Bank Annual Conference on Developmental Economics, 1993.* IBRD/World Bank, Washington, D.C., pp. 335–376.

Norgaard, R.B. (1992) Sustainability and the Economics of Assuring Assets for the Future Generations. Policy Research Working Paper WPS 832. World Bank, Washington, D.C.

North, D.C. (1981) *Structure and Change in Economic History.* W.W. Norton, New York, NY.

North, D.C. (1990) *Institutions, Institutional Change, and Economic Performance.* Cambridge University Press, Cambridge, U.K.

Okun, A.M. (1975) *Equality and Efficiency: The Big Tradeoff.* Brookings Institute, Washington, D.C.

Olson, M., Jr. (1965) *The Logic of Collective Action—Public Goods and the Theory of Groups.* Harvard University Press, Cambridge, MA.

Ostrom, E. (1986) An agenda for the study of institutions. *Public Choice,* 48: 3–25.

Ostrom, E. (1990) Governing the Commons: *The Evolution of Institutions for Collective Action.* Cambridge University Press, Cambridge, U.K.

Ostrom, E. and Walker, J. (1993) Neither markets nor states: linking transformation processes in collective-action arenas. In: *The Handbook of Public Choice.* D. Mueller, Ed. Blackwell, Oxford, U.K.

Ostrom, E., Schroeder, L., and Wynne, S. (1993) *Institutional Incentives and Sustainable Development.* Westview Press, Boulder, CO.

Ostrom, E., Gardner, R., and Walker, J. (1994) *Rules, Games, and Common Pool Resources.* University of Michigan Press, Ann Arbor, MI.

Page, T. (1977) *Conservation and Economic Efficiency: Three Approaches to Materials Policy.* Johns Hopkins University Press, Baltimore, MD.

Pigou, A.C. ([1932] 1960) *The Economics of Welfare,* fourth edition. Macmillan, London, U.K.

Polanyi, K. ([1944] 1957) *The Great Transformation.* Beacon Press, Boston, MA.

Pollack, R.A. (1985) A transaction cost approach to families and households. *Journal of Economic Literature,* 23: 581–608.

Prakash, A. and Gupta, A.K. (1994) Are efficiency, equity, and scale independent? *Ecological Economics*, 10(2): 89–91.

Prakash, A. and Ostrom, E. (1996) Revisiting control, in preparation.

Rawls, J. (1971) *A Theory of Justice.* Belknap Press of Harvard University Press, Cambridge, MA.

Robbins L. ([1932] 1937) *An Essay on the Nature and Significance of Economic Science,* second edition. Macmillan, London, U.K.

Robinson, J. (1979) *Generalizations of the General Theory and Other Essays.* St. Martin's Press, New York, NY.

Schelling, T.C. (1960) *The Strategy of Conflict.* Harvard University Press, Cambridge, MA.

Schumacher, E.F. (1973) *Small Is Beautiful. Economics as If People Mattered.* Harper & Row, New York, NY.

Simon, H. (1957) *Models of Man.* Wiley, New York, NY.

Simon, J.L. (1981) *The Ultimate Resource.* Princeton University Press, Princeton, NJ.

Smith, A. ([1776] 1937) *An Enquiry into the Nature and Causes of Wealth of Nations.* Modern Press, New York, NY.

Stein, A. (1982) Coordination and collaboration: regimes in an anarchic world. *International Organization,* 36: 294–324.

Tacconi, L. and Bennett, J. (1995) Economic implications of intergenerational equity for biodiversity conservation. *Ecological Economics,* 12: 209–223.

Taylor, J.F.A. (1993) The ethical foundations of the market. In: *Rethinking Institutional Analysis and Development.* V. Ostrom, D. Feeny, and H. Picht, Eds. ICS Press, San Francisco, CA, pp. 377–388.

Tinbergen, J. ([1952] 1966) *On the Theory of Economic Policy,* second edition. North-Holland, Amsterdam.

Tocqueville, A. de ([1835] 1945) *Democracy in America.* P. Bradley, Ed. Knopf, New York, NY.

Upadhyaya, R.B. (1976) *Social Responsibility of Business and the Trusteeship Theory of Mahatma Gandhi.* Sterling, New Delhi, India.

Wade, R. (1987) *The Management of Common Property Resources: Finding a Cooperative Solution.* The World Bank, Washington, D.C.

Weimer D.L. and Vining, A.R. (1989) *Policy Analysis: Concept and Practice.* Prentice Hall, Englewood Cliffs, NJ.

Weiss, E.B. (1989) *In Fairness to Future Generations: International Law, Common Patrimony, and Intergenerational Equity.* Ardsley-on-Hudson, New York, NY.

Weiss, E.B. (1990) Our rights and obligations to the future generations for the environment. *American Journal of International Law,* 84(1): 198–207.

Williamson, O.E. (1975) *Market and Hierarchies—Analysis and Antitrust Implications.* Free Press, New York, NY.

Williamson, O.E. (1985) *The Economic Institutions of Capitalism: Firms, Markets, and Relational Contracting.* Free Press, New York, NY.

Wolf, C. (1979) A theory of non-market failures: framework for implementation analysis. *Journal of Law and Economics,* 22: 107–139.

World Commission on Environment and Development (1987) *Our Common Future.* Oxford University Press, Oxford, U.K.

Young, O.R. (1982) *Resource Regimes—National Resource and Social Institutions.* University of California Press, Berkeley, CA.

Young, O.R. (1989) *International Cooperation—Building Regimes for Natural Resources and the Environment.* Cornell University Press, Ithaca, NY.

Sustainable Development: A Southeast Asian Perspective

4

Kiew Bong Heang
Department of Zoology, Faculty of Science, University of Malaya,
Kuala Lumpur, Malaysia

Abstract

This chapter presents some personal insights into the issue of sustainable development with particular reference to Southeast Asia and Malaysia therein. I am of the opinion that it is not practical to strive for environmentally sustainable development. The energy put into trying to attain environmentally sustainable practices is less important than working toward a development that has a wholesome end as its objective, namely, eliminating poverty and human suffering, providing basic infrastructures for all, preventing basic illnesses, providing education and religious guidance, and enabling people to make the most of their opportunities.

I argue that much of the current development in Southeast Asia focuses on at least some of these objectives and, as a result, meets the requirements for sustainable development outlined in the Brundtland report; these requirements, although less stringent than biophysical ones, are more realistic. I discuss the perceptual problems that the North has in relation to development in Southeast Asia and make three main arguments: (1) the region's environmental record is much better than the North's, (2) the North is actually more concerned about increased economic competition from Southeast Asia, and (3) the people of

1-57444-077-2/97/$0.00+$.50
© 1997 by CRC Press LLC

Southeast Asia resent the North's hints of ecocolonialism but welcome international cooperation for the wise use of natural resources. Fundamentally, we believe that we know what we are doing; this belief is derived at least in part from observing the past mistakes of other countries.

Malaysia's development over the past 30 years is summarized, and its prospects for the next 30 years are evaluated. Finally, the provisions of humanity's dominant world views for sustainable development are briefly compared, noting that the largest barrier to achieving any measure of sustainability is a perceptual one.

Introduction

Differences in perception of the term "sustainable development" have often led to different plans and actions for the development of natural resources and the environment. Scientists, economists, sociologists, and politicians all have their respective understanding of the term. This has often led to the formation of a variety of policies, plans, strategies, and programs for the development of natural resources that are often not sustainable in terms of biogeophysical requirements.

Most agendas for development are set for economic growth. Rarely has a sustainable economy been given due attention. Most development projects are designed for profit in the corporate sector. However, development initiated by governments may consider longer term objectives and the needs of improving the quality of life. This development may be more acceptable in terms of social or other kinds of sustainability, but it still may not meet the requirements of biogeophysical sustainability.

These different aims and objectives have often led to differences in perception and understanding of the term "sustainable development." The culture, beliefs, background, and training of an individual influence his perception of things around him, as does his environment. As such, it can be expected that people living in mostly underdeveloped Southeast Asia would have a different perspective on sustainable development from people living in the developed North.

It is necessary for the North to understand the South if the human world is to develop in harmony with the global environment. In this chapter, I relate my understanding and experience of sustainable development in Southeast Asia, with special reference to Malaysia, as a reflection of the South. This understanding comes from a background in natural sciences rather than economics. As a result, the tone of the chapter is as much pragmatic as academic, more a set of opinions than a rigorously structured analysis. It is hoped that this will serve as a refreshing contrast to the more in-depth work that follows. First, the definitional issues of sustainable development are highlighted and then suggestions for sustainable development in Malaysia and neighboring countries are provided. In the process, the differences in perceptions of sustainable development between countries in Southeast Asia and those of the North are revealed.

Southeast Asia as defined here covers the seven member states of the Association of Southeast Asian Nations (ASEAN), namely Indonesia, Malaysia, Thailand, the Philippines, Vietnam, Brunei, and Singapore, and the other non-ASEAN mainland states of Myanmar, Laos, and Cambodia. Papua New Guinea is not included here, but Irian Jaya (the Indonesian part of New Guinea) is.

The Terminology Trap

The term "sustainable development" is a terminology trap. All one has to do to see the controversy is construct the term from definitions in a dictionary. In the *New World Dictionary of the American Language* (Guralnik, 1978), the word "sustainable" is the adjectival form of the verb *sustain,* which is defined as: 1. to keep in existence, keep up, maintain or prolong; 2. to provide for the support of, to provide sustenance or nourishment for; 3. to support from or as from below, carry the weight or burden of; 4. to strengthen the spirits, courage of, comfort, buoy up, encourage; 5. to bear up against, endure, withstand; 6. to undergo or suffer (injury or loss); 7. to uphold the validity or justice of; 8. to confirm, corroborate.

As the definition shows, the verb *sustain* has eight shades of meaning according to its contexts and objectives. The term "sustainable" is defined as that which could be sustained. Likewise, "sustainable" has eight corresponding shades of meaning, which might provide room for potential misunderstanding and conflict when related to "development."

The word "development," in turn, is defined in the same dictionary as: 1. a process of developing or being developed; 2. a step or stage in growth, advancement; 3. an event or happening; 4. a thing that is developed, a number of structures on a large tract of land built by a real-estate developer.

Development as a concept varies from the abstract to a process or action taken, to solid form of perceivable reality. Development is often tied to growth in size, number, or amount. Yet, not all development must have a perceivable increase in quantity. Development could merely take place in qualitative terms, which is less accountable.

To develop is defined as: 1. to cause to grow gradually in some way, cause to become gradually fuller, larger, better, to build up or expand; 2. to make stronger or more effective, strengthen; 3. to bring (something latent or hypothetical) into activity or reality; 4. to cause to unfold or evolve gradually; 5. to make more available or extensive.

The action of developing provides a need for gradual change in form, quantity, quality, and state of existence. The concept of time as it relates to the term "gradual" is also open to interpretation. Sustainable development, which is expected to result in gradual changes to the environment and its resources at a level that can be maintained and prolonged, is a questionable possibility.

Once aware of the variable meanings of the terms and concepts of sustainable development, it is possible to perceive the terminology trap and avoid it. However, avoiding the trap is not what everyone may wish politically. In fact, the trap can be put to good use for selfish ends (e.g., in international negotiations on linkages between trade and the environment).

The Agenda for Development

Southeast Asian nations have their respective agendas for national development and growth. They hope to develop and improve their economies and the standard of living of their citizens. The North, as portrayed in the international mass media, is perceived as a good model to emulate.

The concept of being "developed" is a point of view. The people of Southeast Asia have the misconception from the media that the North has achieved the "developed" state, where most of the people have the desired material wealth, food, and quality of life. Southeast Asians would very much like to have the same, although not all of us subscribe to this misconception. The desired developed state can only be achieved through development of the human resources along with development of land and natural resources to bring about economic growth.

The question to ask is development in relation to what. The Penan in Borneo are developed for their life in a rain forest environment. They are considered undeveloped for city life, a life of industries, trade, or commerce. In terms of technology, the Penan are developed in their own ways and live in harmony with their natural environment. Judging them by the standards of an alien environment makes them only *appear* undeveloped.

Whatever development is, the people of Southeast Asia expect their governments to plan it. They would like to see as much development as possible toward a better quality of life within their lifetime. The countries in Southeast Asia therefore must develop their land and natural resources to bring about economic growth.

To stay in office, politicians in Southeast Asia must to be popular. To secure their tenure in each election, they must come up with development plans that meet the aspirations of the people. Failure to bring development, economic growth, and wealth is a threat to their security. Development cannot be overlooked and must be on every election manifesto. The form of development and whether it is sustainable is not the issue in elections in Southeast Asia. Everyone realizes that development must result in changes, acceptable or otherwise. If development is managed for the improvement of people's quality of life, it is desirable. If development is not catastrophic, it is assumed to be sustainable and is deemed acceptable.

Sustainable development is perceived by the governments of Southeast Asia as an international trade issue linked with such issues as the loss of rain forest and its biodiversity, the depletion of stratospheric ozone, global warming, rising sea level, acid rain, transboundary pollution, and ocean resource depletion. It is not, for the most part, seen in terms of poverty eradication in the Third World, inequality in wealth distribution between North and South, trade imbalance between West and East, and ecocolonialism in the name of environmental conservation.

In development, the governments in Southeast Asia have a number of resources at their disposal that can be utilized for the increase of their gross domestic products (GDPs). Many of the land areas are still in forested wilderness. The forested land can be converted to plantations, which can generate jobs and economic produce. The products of agriculture and forestry can be processed into manufactured goods. These downstream industries will generate more jobs for the people and produce value-added goods for export, which will also boost the GDP.

The use of GDP as an economic indicator does not reflect the concern and interest of individual people. People as individuals are more concerned with their quality of life, which has its own indices: security, freedom, leisure, safe water, clean air, adequate quality food, shelter, infrastructure, facilities, amenities, health care, and education opportunities. "Sustainable" in gross national product or GDP terms is an increase in monetary growth in the coffers of the central bank. It is of interest to the finance minister whose goal appears to be maximizing the power generated by this growth.

In the traditional development model, there are minerals to be mined, oil and gas to be exploited, rivers to be dammed for water or hydroelectric power; fishery resources to be harvested; and aquaculture ponds to be dug for breeding seafood for export to the North. There are also people to be trained and their potential to be enhanced through education. In this model, the mobilization of all the available manpower and natural resources for industries will enable the full corporatization of the nation to work for the improvement of the economy of the country. As long as economic growth can keep up in time, this development can be considered sustainable.

The duration of such economic growth is not a well-received question in political circles. The long-term sustainability of development is not a pressing issue for the politicians at the helms of governments. The term of office of most governments is five years, after which a government has to seek a new mandate.

Talk of a time frame over ten years (two terms of office) is rare and bold. Only in countries with a steady political framework have leaders been able to set longer term agendas toward the year 2020; Malaysia is an example.

Contrary to popular belief in the North, many countries in Southeast Asia still have large areas that could be frontiers for development. The sustainability of the frontier areas is not an issue at the moment. In the last two decades, vast acreages of forested land have been allocated for land schemes and for the transmigration and translocation of people from more crowded areas. This has led to vast areas of forest being cleared away, setting off alarms among environmentalists from the North as to the future of the rain forest and the consequences for the atmosphere. For the people in Southeast Asia, this development is essential for the improvement of their quality of life and the economic progress of their countries.

In response to the growing concern from the North, Southeast Asian environmentalists and consumer groups are also beginning to voice their alarm. However, none of these groups is able to produce acceptable scientific justification for stopping the development.

The land schemes in Malaysia have been success stories, with the nation's oil palm plantation acreage greatly increasing. Some of the transmigration projects in Sumatra were reported to be less successful, with some of the participants from Java returning home after abandoning the scheme started for them. On the whole, the land cleared has been put to use for agriculture of one sort or another.

At the same time, agricultural land has been abandoned, due to changing economic development in the pricing of some agricultural produce. One Malaysian plan was to rehabilitate abandoned agricultural land. The falling price of agricultural commodities, such as rubber, in Malaysia was more a security threat to the country than was the sustainability of the development of the crop. The rural poor, dependent on rubber, were at the point of having their living threatened due to lack of income, and this caused some disturbances in those parts of the country that were severely affected. The security threat led the government to take remedial action in its agricultural policy to ensure the economic sustainability of this industry.

The interest in sustainable development is perceived in Southeast Asia more as a trade issue created by environmentalists in the North to provide advantage for their domestic interests than for the integrity of the Southeast Asian environment or the sustainability of its development. The price fluctuation of basic commodities from the region is determined in the North, out of the control of the people affected in Southeast Asia. The non-sustainability of the prices of commodities produced in Southeast Asia is more a threat to the stability of the region and the sustainability of agricultural practices than to the environment, especially ecological processes. This destabilizing factor has rarely been seen by the governments of Southeast Asia to be of concern to environmentalists from the North.

On the contrary, environmental and sustainability issues have often been coupled with human rights issues in hindering development in Southeast Asia. Objections to the logging of native land in Sarawak have attracted a considerable amount of attention in the media. The blockage set up by the Penan in Sarawak against loggers was claimed by the Malaysian authorities to be instigated by an environmentalist from the North.

Environmental groups from the North even send their members to chain themselves to tractors and cranes as a sign of their protest. The low price of tropical hardwoods in the international market was claimed by Malaysian politicians to be one of the causes of overlogging. As much as the North desires, countries logging their forest for commercial gain would also like to reduce the cutting of their forest as well. If the timber were more valuable, the incentive to conserve it would be much increased.

However, an established industry cannot be terminated without social problems for the people involved if time is not provided for the establishment of sustainable alternatives. The availability of sustainable alternatives is not known locally. Neither have the objectors to logging come forward with alternative economic options that can be adopted to preserve the rain forest.

Alternative logging practices, such as the use of helicopters, enable logs to be cut from otherwise quite impossible terrain. Portable sawmills, which require a team of four to six people, have been promoted in Papua New Guinea; they have also been promoted in Sarawak for the extraction of Bornean iron wood, but at a scale that does not yield enough timber to make this wood a major export commodity. Such a scale for the logging industry would ensure a measure of ecological sustainability but not economic sustainability and as such is not a viable option.

In light of these problems, the prime minister of Malaysia called on the international community for compensation to preserve the country's rain forests but did not receive any offers from the North. Even the debt swap idea proposed in South America has not been offered as a sign of willingness and sincerity.

In addition to logging, the construction of hydroelectric dams, which also result in the loss of rain forest, has received concern from some in the North. The nations of Southeast Asia are not unaware of their loss, but they see the development of hydroelectric dams as an energy option for their countries in the face of rising oil prices from OPEC countries. The sacrifice of the forest for electricity is needed to improve the quality of life and for industrial and urban development. It is a conscious choice that has been made as the result of a trade-off. The opinion of many Malaysians, including myself, is that the land and the forest belong to the people who make the choice. Outsiders do not have the right to interfere without payment for such a right.

The people of Southeast Asia at times perceive the influence from the North as a tactical suppression of the development of the countries in Southeast Asia in the name of human rights and environmental issues. The North may be worried about the increased competitiveness in parts of Southeast Asia that would result from the development of industries and natural resources.

The success of the Malaysian oil palm industry in the international market, with its high production of palm oil replacing soya bean oil in many industrial processes and products, has threatened the less economically sustainable soya bean industry in the United States. The Soya Bean Producer Association started a campaign to discredit palm oil, using statistics about cholesterol to frighten consumers in the United States (according to press reports in Malaysia). The campaign was initially successful, costing Malaysia a great deal and forcing the Malaysian government to initiate a countercampaign and provide facts to rectify the damage.

The palm oil campaign was perceived as an unfair trade practice by the Malaysian government. Likewise, the call to boycott tropical hardwood in the North is considered in Southeast Asia to have nothing to do with sustainability; it is linked to vested interests of the Northern softwood industries.

The governments in Southeast Asia believe they know what they are doing. The authorities proclaim that they love their countries and have no intention of destroying their motherland. They believe that the developments they have planned for and put into action are economically and environmentally sustainable.

It is clear from the proclamations of politicians that they view sustainable development as chiefly concerned with sustained economic growth and political stability. Sustainable development is equal to a sustainable economy and has nothing to do with the sustainability of the natural biogeophysical foundation as scientists see it.

The agenda set for most Southeast Asian nations addresses the development of their human resources, along with their land and natural resources. Most Southeast Asian nations have developed their agriculture, forestry, and fisheries sectors. They hope to use these industries as a springboard for industrialization. The service sector (tourism, trade, and commerce) is also growing strongly. The aspiration is to be as "developed" as possible.

Constraints of Sustainable Development in the Region

The main biogeophysical constraint for development in most parts of the Southeast Asia is the high risk of erosion and leaching of the soil due to heavy rain. The monsoon areas on the mainland suffer periodic droughts, which restrict plant growth and crop productivity. The rainy season may bring floods in agricultural regions, damaging crops. The lack of fertility of the land is also a constraint for agricultural development in some areas. Certain crops grow best in certain soils. Soil surveys need to be carried out before an agricultural program can be implemented for the selection of a suitable crop.

Some areas in Southeast Asia are subjected to natural disasters, such as drought, flooding, typhoons, earthquakes, and volcanic eruptions. In these high-risk areas, infrastructures may be destroyed so frequently that sustainable development may not be a practical consideration. The development and successful use of these areas for economic progress depends to some extent on luck and timing. The success of a development program can only be sustained over a short period of time. Long-term sustainability is often out of the question without productive economic yield from the land over a short period of time, such as the planting of annual crops.

In addition to these biogeophysical constraints, other constraints are chiefly socioeconomical in nature. Parts of mainland Southeast Asia, the Philippines, and Indonesia have security problems related to insurgencies and civil wars. Development in these parts is a high-risk enterprise and is not generally undertaken by investors. These areas remain frontiers for development.

One of the legacies of the colonial powers is an administrative infrastructure for development. Some of these legacies can be considered an asset, while many others are a hindrance that served only the interests of the colonial masters. Many of the laws and regulations left by the colonial authorities were designed for the exploitation of the resources for the good of the foreign countries, which may not have long-term objectives. Sustainable development, by definition, needs long-term objectives, which those laws in particular do not provide.

Toward Sustainable Development in Southeast Asia

International Perspective

The call for sustainable development was initiated by the developed countries in the North. These countries played a major role in producing the Brundtland Commission's definition of sustainable development (IUCN-UNEP-WWF, 1980) as development that meets the needs of the present without compromising the ability of future generations to meet their own needs. I argue that development in Southeast Asia at present is able to satisfy the terms of the Brundtland Commission's definition.

There have been complaints from countries in the North that some countries in Southeast Asia are overlogging their forests. However, countries like Malaysia and Indonesia felt that there was enough forest to meet the needs of the present without compromising the ability of future generations to meet their own needs. In these countries, the option of future generations to use the gene pool of the forest is sufficiently provided for in the national parks, forest reserves, and game reserves set aside for conservation and preservation. Research has been initiated to develop the biomedical potential of the species inhabiting the forest and the sea.

There is a considerable amount of deforestation in Southeast Asia; that much is certain. Countries such as Thailand, the Philippines, and parts of Indonesia have little forest left. However, other parts of Southeast Asia still have good forest cover. The area is vast, even though not very much of it is virgin forest now. However, the total preservation of vast tracts of forest would deny the use of forest resources and would therefore not be deemed acceptable. Countries such as Malaysia expect to preserve about 10% of the land under protected forest for preservation in the form of national parks and virgin forest reserves, 40% of the land under forest estates targeted for management on a sustained yield basis, and another 30% under tree crops, such as oil palm and rubber. Such a formula is highly respectable by international standards for the use of land in an ecologically acceptable state. The total abstinence of logging parts of forest estates would be deemed foolish and a waste of potential economic gain that could be realized now while the market is good. The forest is also capable of regeneration if managed properly.

In 1991, a new definition for sustainable development was proposed by the International Union for the Conservation of Nature (IUCN), the United Nations Environment Program (UNEP) and the World Wildlife Fund (WWF), along with a definition of a sustainable economy. Sustainable development was defined as improving the quality of human life while living within the carrying capacity of supporting ecosystems.

In Southeast Asia, development exceeding the carrying capacity of supporting ecosystems is not visible to the people. To most of the citizens of Southeast Asia, there is no such problem in their region. Millions of people in Singapore are able to live an enviable quality of life in the cramped metropolitan area of the island state. Their standard of living resembles that attained by the developed North.

If there is an issue of concern in Southeast Asia, it is the standard of living of rural people. Many rural areas are without running water, electricity, telecommunications, and other present-day amenities expected in the North. These are the areas that require the development of infrastructures, facilities, amenities, and services that are already available in many of the cities and in the developed North. The development and provision of these goods would contribute tremendously to uplifting the quality of human life in these areas. If the development of these goods is sustainable for the cities and the developed North, can their development be considered not sustainable for these rural people? This is a question that needs to be answered.

The possibility of development exceeding the earth's carrying capacity is not visible at present, but the difference in the quality of life between North and South is clear. Improvements in the quality of life brought about by the development of better infrastructures, facilities, amenities, and services seems a good idea to the people of Southeast Asia who do not have these things. Therefore, to raise the question of long-term sustainability seems rather inappropriate.

The definition proposed by IUCN-UNEP-WWF (1991) for a sustainable economy, as distinct from environmental sustainability, appears more pragmatic. It is essentially the maintenance of the resource base in development to meet people's aspirations. This seems possible in Southeast Asia with the potential of the people and nations in the region. To adapt through improvements in knowledge, organization, technical efficiency, and wisdom to the changes brought about by development is within their control.

If a sustainable economy is attainable, it would follow that sustainable development is possible. What remains to be formulated is an action plan for improving the quality of life of the people in the region. The development and implementation of such a plan should, in my view, be by the people for the people. It is their life and destiny at stake. The decision is theirs to make.

The attainment of biogeophysical sustainability is quite an impossible issue in Southeast Asia. Biogeophysical sustainability requires the perpetuation of the character and natural processes of ecosystems, and the indefinite maintenance of their integrity as productive, diverse, stable, and adaptable units, without degrading the integrity of other ecosystems. Such a development concept is an impossibility and does not make any sense to local people in Southeast Asia who do not wish to remain in their present undeveloped state.

In today's fast-changing world, an action in one part of the world could have effects in another. This is a valid issue of concern for the developed North, as evidenced by the greenhouse effect, polar holes in the ozone layer, rising sea level, and general pollution. However, these problems should not be used to hamper the development of Southeast Asia, because the region's contribution to such global environmental problems is small, if not insignificant. The North has much to do to rectify its behavior without the nations in Southeast Asia having to sacrifice their development. Payment for the damage done to the world by global pollution must by properly allocated. If this principle is followed, Southeast Asia should be allowed to log its forests for timber and open up more land for agriculture and other development.

Perhaps these views seem foolish. True enough, the people of Southeast Asia are not all wise. The reduction of the available carbon sink in the conversion of forests to tree crop plantations *appears* to be a reasonable trade-off in light of the difficulty of measuring reductions in carbon dioxide absorption (it requires the use of highly sensitive instruments). The logging of the forest enables the carbon collected in the logs to be utilized while the logged-over forest continues to function as a carbon sink at a more efficient level in its regeneration. Most climax rain forests show a stochastic balance of photosynthesis and respiration. The net capture of carbon from the atmosphere may be small and insignificant in a climax forest compared with a tree-crop plantation and a regenerating forest, both of which show visible increments of growth in overall biomass.

Ultimately, the people of Southeast Asia feel that if they do not intercede with development in the North, then why should the North interfere with their development? Of course, international cooperation for the improvement of the world is acceptable, but all parties must share appropriate proportions of responsibility. In relation to logging, for example, this cooperation might express itself in the establishment of an international agency modeled along the lines of the Convention on International Trade in Endangered Species (CITES). Such an agency should also be able to regulate harvests, pricing, and marketing. However, any attempt at neocolonialism is not welcome in Southeast Asia, especially with its past history of colonialism.

Southeast Asian Perspective

In the prehistorical past, the aboriginal people of Southeast Asia were invaded and taken over by more aggressive neighbors from established kingdoms, such as India and China. New kingdoms were formed that had their seats of government mainly along the coast. Here, the aboriginal people, living a hunter-gatherer and horticultural lifestyle, were replaced by people with a self-sufficient agricultural lifestyle. Both these lifestyles, at their respective population levels, were environmentally sustainable.

Later, the Western colonial powers arrived, along with economic immigrants from China and India, who established themselves along the coast, setting up towns and cities. These people brought in trade and commerce and introduced mining to the region. They also introduced commercial agriculture for the production of crops for export. The result was a larger scale of destruction to the environment in the area. The forests were logged for export and the land was opened up for plantations.

The colonial powers and economic immigrants came to Southeast Asia to make money from mining and trading and then to return home. They had little or no loyalty to the land. Their basic objective was to generate wealth at the expense of the land and its natural resources.

The Southeast Asian region was fortunate to have tree-crop plantations, such as rubber and oil palm, developed on the land. These are more appropriate in the tropical rain forest environment than pasture for cattle ranching. The tree crops are acceptable substitutes for the replacement of rain forest trees destroyed. The region would have suffered greater environmental degradation if the rain forest had been opened up for pastures.

The development of commercial agriculture was followed by industrialization, which made use of the local human resources. At present, industrialization appears to be sustainable, as it uses the products of the land and puts added value on the products. Some industrialization also involves the manufacturing of electronic products and garments. These developments, along with the development of human resources, appear economically sustainable so far in Malaysia, with more than eight years of continuous economic growth.

However, many industries produce pollution due to a lack of concern for the environment. Resource utilization appears sustainable from the available production of the land but is not sustainable with regard to the pollution generated. If the pollution problem can be resolved, the industrial sector would make a sizeable jump toward environmental sustainability.

The operation and process of trading provide a means of adding value to the products of the land through the management of distribution. The transportation industries associated with trade may not be sustainable, as they are dependent on fossil fuels, but the process of exchange itself does no harm to the environment and can therefore be encouraged.

Southeast Asia, on the whole, is not in a state of catastrophe. Life in Southeast Asia is acceptable in all areas except those in a state of war. The quality of life in many rural areas has room for improvement with the provision of infrastructure, but at least there is no known starvation. The situation looks sustainable in its present state despite pockets of environmental degradation which need remedial action.

The Malaysian Perspective

In Malaysia, national development is scheduled in five-year plans. The First Malaysia Plan (1966–1970) took stock of the nation's resources and identified the problems and potential for development.

The Second Malaysia Plan (1971–1975) set the agenda for the utilization of the land and resources for national development and initiated the process. The Third Malaysia Plan (1976–1980) produced reviews of the development achieved and provided for environmental protection to balance the benefit obtained with the deterioration caused to the environment. The Fourth Malaysia Plan (1981–1985) provided new directions for enhancing national development and set the agenda for accelerating the industrialization process. The Fifth and Sixth Malaysia Plans (1986–1995) proceeded with the industrialization of the country while developing the land-based natural resources. The net result is impressive economic growth over the duration of these plans, giving the impression, at least among much of the populace, that the country has struck the right formula for a sustainable economy.

With the increase in industrialization, efforts must be made to establish markets for the products generated. This calls for the internationalization of trade, along with a more active part in international politics. Local cooperations were encouraged to set up trading bases and investments overseas. There was also an attempt to organize the nation as a "corporate unit" (Lim, 1995). The government of Malaysia recognizes the need to be proactive in the international arena in politics, economics, and environmental issues. It has now become a leader among the countries in the South and a spokesperson for the Third World.

In 1982, Malaysia started a Look East Policy to focus the nation on the examples set by Japan and free itself of the binding influence of the West. The nation has successfully followed the paths of Korea and Taiwan in their economic growth and has emerged as a newly developing nation in the region, after Singapore. In recognition of its limitation in technology, the Malaysian government has had the wisdom to invite foreign investments and the technologies that accompany them. These investments have spawned such Malaysian industries as automobiles with the help of Japan. Malaysia is a leading producer of rubber, palm oil, and timber. It also has a well-developed oil and gas production industry and is one of the leading producers of microchips for the electronics industry around the world.

Recognizing the limitation of the markets in the North, the country is actively pursuing the establishment of new markets among the nations of the South, in particular the former Soviet Union, China, the Middle East, India, and Africa. The country seeks to internationalize its trade and commerce while adding value to its products from agriculture and other natural resources under exploitation, such as timber and minerals.

In the course of development, the country has not been spared the problem of pollution. The Third Malaysia Plan provided for the founding of the Department of Environment. The country at present has good protective legislation for the environment, comparable to developed countries. It is also actively pursuing the development of appropriate waste-disposal technology.

In the protection of the environment, industrial development was allowed to continue with the introduction of a graded standard for waste discharge in stages. This procedure has successfully dealt with pollution from the agronomic industries (e.g., rubber and oil palm processing). The nation is still struggling with hazardous waste from many of its industries and is in the process of setting up a hazardous-waste treatment plant.

The country has a sound environmental policy that maintains over 58% of the land under forest cover. Taking the rubber and oil palm plantations into consideration, the country has well over 87% of its land area under tree cover, which is better than most countries in the developed North.

The country has mixed energy policies which balance the use of coal, oil, gas, and hydropower for electric generation. It has stayed away from the use of nuclear power in recognition of the hazard of nuclear waste. There are plans in every Malaysian state to ensure the availability of sufficient potable water to the year 2050. All water catchment areas have been identified, and protection of their integrity has been given due importance. The protection of these catchment areas will contribute to the further protection of Malaysia's biodiversity and the ecological integrity of the land.

Fortunately for Malaysia, the country has a stable political system in operation, and the leader of the government has a vision for the future. The prime minister has set a new goal for the country, called Vision 2020, which targets the nation attaining developed status by the year 2020. The political stability of the country with the end of the Communist insurgence is the main condition permitting the impressive economic growth and development. The security of the country without "war zone" areas allows for better distribution of development and use of the available land and natural resources.

Can such economic growth be sustained until 2020? The indication so far is positive. Is such growth sustainable in terms of biogeophysical sustainability? The answer is most likely not. The aspirations of Malaysians can be shared by all the nations of the South. The pattern and process of development can be emulated for the development of underdeveloped nations around the world. The success of Malaysia's development can do all Malaysians proud. Of course, such effusiveness may make Malaysia seem like a paradise, but it *is* a paradise in my perspective, in the sense that I would not consider living anywhere else. It is sad to say that not everyone living in paradise appreciates it.

Conclusions and Recommendations

The belief system of a person leads to the type of action he takes, which produces the consequences that follow (Berry and Horton, 1974). There is a need to garner the healthy principles of religions in the responsible use of natural resources and the environment and emphasize these principles in places of worship.

The laws of classical economics are obeyed with devotion, resulting in the conservative economic stance of the world today, the outcome of which is the exploitative attitude toward the environment and its resources, which are seen as unlimited subsets of the economy. There is a need to change the ways the economy of the world is practiced, as suggested by the new discipline of eco-logical economics: recognizing the limits of the environment and working on the basis that the human economy is a subset of the environment and not the other way around.

The Judeo-Christian and Islamic religions conceive that man is superior to all other creatures. Man is endowed with a God-given right to use, modify, and exploit the environment as he sees fit, using whatever power he is given, al-though he also has the responsibility of stewardship. The Buddhist religion sees man as an entity given the opportunity to attain Nirvana and all life as of equal significance in this world.

Animalistic religion, by contrast, perceives man as a part of the environment under a trusted Guardian power from whom permission must be sought in the use of resources. Wanton destruction of environmental resources will bring forth the wrath of the Guardian with undesirable consequences. Primitive societies with this religious belief system appear to maintain a sustainable lifestyle with due respect to the environment and its natural resources. The Buddhist belief tends to protect sentient life forms, such as animals, which are believed to have a mind entity capable of attaining buddha-hood, but the religion fails to denote sufficient protection to plant life and the environment as a whole. The Judeo-Christian and Muslim beliefs do not require much protection for other life forms and the rest of the natural environment, as their emphasis is anthropocentric.

Urban society and its economy, remote from nature, worship money and the power associated with it. The exploitation of the environment for a person's creature comforts is perceived to be just and right. Any suffering caused outside a person's perceivable environment is not his concern, as it does not affect him directly or his immediate world of bricks and concrete.

The communist system, which explicitly recognizes the public interest above sectarian interests, should be able to protect the environment and its resources (Berry and Horton, 1974), but observation of many socialist states shows this not to be so. The communist system so far has not proven itself to be better than the capitalist system, as development under both systems has stressed the need to generate profitable output, which tends to lead to the expression of greed.

The common property of the environment, which belongs to everyone and no one in particular, has resulted in its resources being treated as free and of no value in the accounting system. Under the current accounting system, environ-mental sustainability is meaningless without profit in the financial world, which governs the world's economy.

The beliefs and perceptions of people do not allow them to develop their environment in a sustainable manner with reference to principles of biogeophysical sustainability. Sustainable development, in terms of ensuring the preservation of the elements and operating processes of the biogeophysical world, is not practical, given the present stage of human evolution, which stresses the use of technology to attain power over nature in an effort to secure control over the potential risks caused by nature.

Economic interests will continue to dominate the world, with economists acting as high priests in the temples of political power. Such an undesirable state will perpetuate itself as long as ecologists and other natural scientists are unwilling to become deeply involved in politics and play a more proactive role in the governance of nations.

Unless there is a change in the perception of our role in nature and of the limitations of the biosphere, sustainable development will remain a dream. The future of mankind is predicted to be bleak when we exceed the carrying capacity of the world, resulting in widespread catastrophe and misery to all those then alive. Southeast Asians have as much a role to play in averting that catastrophe as those in the North and elsewhere.

References

Berry, B.J.L. and Horton, F.E., Eds. (1974) *Urban Environmental Management: Planning for Pollution Control.* Prentice-Hall, Englewood Cliffs, NJ.

Guralnik, D.B., Ed. (1978) *New World Dictionary of the American Language, Second College Edition.* William Collins & World Publishing Co., Tulsa, OK.

IUCN-UNEP-WWF (1980) *World Conservation Strategy: Living Resource Conservation for Sustainable Development.* International Union for the Conservation of Nature, in conjunction with United Nations Environment Programme and World Wildlife Fund, Gland, Switzerland.

IUCN-UNEP-WWF (1991) *Caring for the Earth: A Strategy for Sustainable Living.* International Union for the Conservation of Nature, in conjunction with United Nations Environment Programme and World Wildlife Fund, Gland, Switzerland.

Lim, K.W. (1995) *Malaysia Incorporated: An Emerging Asian Economic Powerhouse.* Limkokwing Integrated Sdn Bhd., Kuala Lumpur, Malaysia.

World Commission on Environment and Development (1987) *Our Common Future.* Oxford University Press, Oxford, U.K.

Practical Steps Toward Sustainability

Part II

Economic vs. Financial Pricing of Timber and Its Probable Impact on National Accounts: The Costa Rican Case, 1980-92

5

Juan Antonio Aguirre Gonzalez
Centro Agronómico Tropical para Investigaciónes y Enseñanza (CATIE),
Turrialba, Costa Rica

Abstract

The use of financial prices has underestimated the importance of the forestry sector in Costa Rica's economy. Using data from 1980 to 1991, the contribution of the forestry sector to total GDP ranges between 11.5 and 24.5% using economic prices, but only 3 to 5.4% using financial prices. The use of financial prices has resulted in negative nominal protection rates and high subsidy levels. The consequences of this undervaluation are (1) little interest in caring for the forest, (2) a weak voice for the forestry sector in the general policy dialogue, and (3) very low economic returns to producers. The result has been massive deforestation in Costa Rica during the 1980s, with its attendant environmental problems, and little interest in sustainable forest management.

By contrast, when a variety of uses of tropical forest are considered in addition to timber (e.g., carbon fixation capabilities, water for energy production

1-57444-077-2/97/$0.00+$.50
© 1997 by CRC Press LLC

and domestic use, pharmaceuticals, food), the value of a hectare of forest is nearly $900 per year in 1990 U.S. currency. This value is far greater than the value of timber alone, but without more representative market prices, it cannot be translated into sustainable use. Nonetheless, better market prices mean more money for the people who use the forest directly. This is a language that all farmers can understand, regardless of the scale or type of their operations or their ethnic and national origins. Better prices now will buy the time to develop the sustainable practices and knowledge needed for long-term conservation of tropical forests, in Costa Rica and elsewhere.

Introduction

For a long time now, developing countries have practiced timber-pricing policies that undervalue their tropical timber resources. Inevitably, such policies will come to threaten the long-term survival of the sector. They encourage waste, industrial inefficiency, and an absolute disregard for sustainable management of tropical timber resources. Costa Rica is clearly no exception to such domestic pricing policies.

Because of this threat to the sector, natural resource economists have focused much of their attention in recent times on developing more accurate methods for assessing the value of tropical rain forest resources. They have devoted efforts to identifying non-timber forest products, estimating their value correctly, and incorporating other environmental externalities. When all this is added to the value of extracted timber, the result will presumably give the true total value of tropical forest production in terms of costs and benefits. Recent studies clearly point in this direction (McNeely, 1988; Panayotou and Ashton, 1992; Panayotou, 1993; Godoy and Bawa, 1993; Godoy et al., 1993; Muul, 1994).

Many ecological economists defend this comprehensive approach to assessing the full value of the forest, as a response to the problem of the low selling value for marketable products of tropical rain forests. This is based on the popular argument that classical economic thought has very little to contribute to the kind of analysis oriented toward ecological economics.

The contention here is that traditional price analysis has in fact much to offer to ecological economics and that traditional economic analysis is indeed relevant to the issue of sustainable management of the tropical rain forest. The destruction of rain forests over the past quarter century can be attributed in good measure to "perverse" pricing policies at the local level.

Ecological economists have generally tended to blame a number of environmentally related issues for destroying the tropical rain forest ecosystem. They have downplayed or outright overlooked the impact of "appropriate" prices as the most powerful motivator of economic activities whenever normal human beings are faced with decisions.

By contrast, forest economists and agricultural economists feel that traditional price policies in developing countries have worked against forest production, instead favoring agricultural and livestock production activities in the region. Such price distortions, they feel, have decimated the area's natural resource base, including the forest.

These positions have come increasingly into conflict, and it can no longer be denied that traditional economics and ecological economics have failed to blend their views into a useful new body of analysis. Only such a composite analysis can possibly provide powerful clues as to why Latin American tropical forests have continued to suffer from relentless wasteful destruction over the past 50 years and why the full contribution of forests to national income has been so profoundly underestimated (CCT/WRI, 1991).

Traditional price policies have consistently undervalued forest resources and have provided an erroneous picture of the real importance of the forest sector and its contribution to the economies of tropical nations. Such perverse policies have systematically undermined the sector's ability to play the economic and environmental role it certainly deserves.

This chapter will argue that there is a pressing need for better pricing systems on timber resources and more complete knowledge of price trends for major timber products. This need exists regardless of whether externalities are handled properly or non-timber products from the tropical forest are identified and accounted for. More accurate pricing systems and more complete information are the only ways to guarantee better resource allocation among the members of the region's traditional agricultural sector, which normally includes the forest sector.

If timber and other natural goods and services from the forest are priced appropriately, the foundations will be in place for a major policy dialogue among sectors. The forestry sector will finally play the economic role it truly deserves, because the real dimension of its contributions will be understood, accounted for, and taken into consideration when major economic policy decisions are made.

Economic pricing means that timber resources are valued at their economic, shadow, or border price. This corrects the imperfection traditionally found in national income accounting, based as it is on a market price. Also known as the financial price, the figure used in national accounts is normally well below economic world-market prices for timber due to multiple local imperfections and is subject to major distortions. This traditional method never really reveals or arrives at the "true" economic value of the tropical rain forest's total contribution to society. Such distortions need to be eliminated or reduced significantly.

The objectives of the study presented in this chapter are:

1. To evaluate the general trends of relative reported world timber prices and compare them with domestic price movements, examining the impact of these price trends on local economic prices of timber over the long term
2. To ascertain how economic pricing affects the value of total timber production and gauge the implications of this pricing method for forest sector accounts, gross agricultural output, and the overall book value of forested lands and their contribution to Costa Rican society

The study is based on two general hypotheses:

1. The use of economic prices instead of financial prices for assessing the value of timber resources in the tropical rain forest has been very detrimental because it understates the role of the forest sector as a share of the total agricultural sector and national accounts, as well as the total estimated value of acreage in tropical rain forests.
2. Relative prices, nominal protection rates, levels of subsidies, and intersectoral transfers all serve as good indicators of the negative impact that traditional price policies have had on the underdeveloped tropical countries and are useful for measuring the perversity of such price policies.

Methods

The study consisted of four major steps. The first step was to estimate relative timber prices and probable long-term price trends. A price index was developed and analyzed, taking 1990 as the base year, to evaluate the behavior of relative world prices for timber and compare them to prices of major tropical commodities. The second step was to gauge the economic prices of timber, as well as their overall impact on the forest sector. Nominal protection rates and levels of subsidies and transfers were estimated, and the results were used to recalculate sector accounts based on economic prices and total intersectoral transfers. The third step was to estimate the total macroeconomic value of the tropical rain forest in Costa Rica. Major timber and non-timber products were included, along with the financial and economic prices of average production from the tropical rain forest and a preliminary estimate of the macroeconomic value of tropical forest acreage. In the fourth step, the results of each of these three analytical processes were woven into a draft policy framework for forests. It was thus possible to judge the major implications that economic pricing and accurate valuation of tropical forest acreage will have in designing a forest policy that can make a positive contribution to sustainable use of tropical forest resources.

Results

Relative Price Behavior and Foreseeable Trends

The concept of relative prices has been widely used by economists for many years to justify resource allocation practices that favor a given commodity and castigate others. This system has fueled the unrelenting standoff between environmentalists and livestock producers.

Environmentalists attribute deforestation to four kinds of policy supports that have been lavished upon the livestock sector:

- International credit support from multilateral and bilateral agencies, with the understanding that land improvement included removal of the forest to establish pastures
- Subsidized interest rates, which have often favored agricultural and livestock activities over the forest sector and allocated short-term resources to activities amply proven to lack long-term economic viability
- Land settlement policies that promoted the occupation and development of lands suited only for forest cover, while the best farm land generally remained in the hands of large landholders, and the true production potential of frontier lands was overlooked
- Agricultural development policies based on the assumption that the agricultural frontier was essentially inexhaustible

Environmentalists have been correct in wielding these arguments. In fact, in some countries, such practices continue to be a part of current agricultural policies. However, an additional argument needs to be included, namely that relative prices, by castigating timber products in favor of other tropical products, have depressed productivity in areas that are already developed, accelerating the expansion of the agricultural frontier into areas clearly suited for forest use only.

The first four arguments have been widely covered in the environmental literature and to a great extent, although indirectly, in the newly emerging material on ecological economics. However, a search of the literature reveals that the relative price argument has hardly been used at all to explain the ecological degradation of tropical rain forests today. Despite the paucity of attention it has received, the relative price argument might actually have been the accelerator driving the rapid transition from forest to pasture that has swept the tropical world, particularly in Costa Rica.

Table 5.1 Past and Projected Future Commodity Prices (US$) and Relative Prices (in parentheses) in Costa Rica, 1970–2005

Year	Beef (¢/kg)	Bananas ($/tonne)	African palm oil ($/tonne)	Logs, meranti ($/m³)	Logs, sapelli ($/m³)	Sawn wood ($/m³)
1970	520 (203)	659 (122)	1037 (358)	148 (84)	171 (50)	370 (71)
1980	384 (150)	527 (97)	811 (280)	271 (153)	350 (102)	507 (97)
1985	314 (123)	551 (102)	730 (252)	199 (112)	253 (74)	403 (77)
1989	271 (106)	578 (107)	370 (128)	201 (114)	289 (84)	446 (85)
1990	256 (100)	541 (100)	290 (100)	177 (100)	344 (100)	524 (100)
1991	260 (102)	547 (101)	332 (114)	196 (111)	309 (90)	462 (88)
1992	230 (90)	444 (82)	369 (127)	196 (111)	311 (90)	481 (92)
1993	244 (95)	413 (76)	352 (121)	364 (206)	290 (84)	502 (96)
1994	235 (92)	408 (75)	358 (123)	280 (158)	293 (85)	514 (98)
1995	231 (90)	397 (73)	334 (115)	289 (163)	298 (87)	517 (99)
1996	228 (89)	426 (79)	307 (106)	298 (168)	302 (88)	523 (100)
2000	274 (107)	409 (76)	304 (105)	303 (171)	326 (95)	545 (104)
2005	262 (102)	401 (74)	267 (92)	320 (181)	361 (105)	572 (109)

Note: Prepared by the author from World Bank data. Base year 1990.

Table 5.1 shows absolute and relative price trends over time for some of the major tropical commodities. It is clear that until the early 1980s, timber maintained relatively lower prices than the other commodities. The relative prices given in Table 5.2 reinforce this perception and confirm our suspicions.

The argument needs to be even further refined. The price of timber is reported in three categories: meranti logs, sapelli logs and sawn wood. These relative prices can be calculated and matched against the weighted relative price indices of coffee, cocoa, sugar, beef, bananas, oranges, and African palm, which continue to compete with timber for land use in the American tropics. Such a comparison shows that the relationship has changed, with prices now moving in favor of timber. The trend toward timber and against crops and livestock can be expected to continue indefinitely (see Table 5.2).

World Bank data paint a clear picture for the coming decade. If expectations are borne out, sustainable exploitation of the tropical rain forest will certainly become a very attractive option for individual investors and for society overall. This global movement will probably extend to Costa Rica, where the very evident trend all over the country is to abandon pasture lands to their biological destiny, secondary forest.

Such new developments are already in evidence in tropical rain forest areas

Table 5.2   Relative Prices of Three Types of Timber (Meranti [M], Sapelli [S], Sawn Wood [sw]) Compared with Other Major Commodities (Beef, Bananas, African Palm Oil), 1970–2005

Year	Beef vs.			Bananas vs.			African palm oil vs.		
	M	S	sw	M	S	sw	M	S	sw
1970	243	409	288	146	245	173	428	719	506
1980	98	147	155	64	96	101	183	275	289
1985	109	167	159	91	138	132	224	342	327
1989	93	126	124	94	127	126	112	152	150
1990	100	100	100	100	100	100	100	100	100
1991	92	113	115	91	113	115	103	127	130
1992	81	99	98	74	91	89	115	141	127
1993	46	113	99	37	91	80	59	144	127
1994	58	108	94	48	89	77	78	145	126
1995	55	104	91	45	85	74	71	133	117
1996	53	101	89	47	90	79	63	121	106
2000	63	113	103	44	80	73	61	111	101
2005	57	98	94	41	71	68	51	88	84

Note: Prepared by the author from World Bank data. Base year 1990.

on the Atlantic coast, where major livestock production activities have flourished over the past 25 years. A survey on the issue was recently conducted by the author and, although findings are still being processed, preliminary figures indicate that extensive tropical beef ranching is being abandoned in favor of "secondary forest growth." Equally significant, out of a sample of 40 farms that still have substantial forest resources, nearly 84% of those interviewed claimed that had they foreseen the collapse of beef prices, they would have left intact at least 40 to 60% of the land they now have under pasture.

Another clear sign of the back-transition, which sees pasture reverting to forest, is the outcry among Costa Rican beef producers in recent months, demanding full-scale economic support in the form of subsidized credit and other inputs and technical assistance. These subsidies had passed into history on the heels of structural adjustment negotiations with the World Bank and the International Development Bank, but producer groups now claim they are necessary for the survival of livestock activities. Incidentally, these are the same policies that were loudly acclaimed in the past.

Findings coming in from current surveys indeed confirm this trend, with relative prices moving against the livestock sector. Local beef producers are beginning to feel the price pinch and are seriously considering abandoning the activity, at least on less profitable marginal lands. Given present world beef price conditions, as well as absolute and relative price prospects for the foreseeable long term, the production of beef and some other traditional tropical products has ceased to be attractive.

With world prices for the different forms of timber on the rise, economic, shadow, or border prices for these commodities will almost certainly maintain their current high levels. There is even a very strong possibility that they will continue to climb, pushing up local timber prices (Aguirre, 1993).

Supply shortages now seem to loom on the horizon, as demand for timber attempts to keep pace with the needs of a growing population and rising income levels in tropical timber-producing countries and the developed world. A recent study placed population growth and a rising per-capita income base on a partial equilibrium model with the demand for timber in Costa Rica. If the results are to be believed, future demand for hardwood in the country will be a direct function of the timber needs of this ever-larger population.

Domestic prices will soon have to be adjusted to reflect a rising world price for timber and the population and income growth being experienced by Costa Rica. This will definitely improve the long-term position of the forest sector in the local economy, particularly boosting the profitability of investments in sustainable forest production. Indeed, at least a half-dozen new non-governmental organizations in the area now offer technical assistance in sustainable management of the humid and dry tropical rain forest.

Impact of Economic Prices in Macro Accounts of the Forestry Subsector and Implications for Agricultural Sector Accounts

Costa Rican timber has traditionally been sold as a cheap commodity by comparison with world timber. A recent study conducted at CATIE (Aguirre, 1993) clearly showed that even though prices for local industrial-use logs had risen sharply in the past five years, they were still far below comparable world prices.

One indicator that domestic timber prices are indeed on the rise is the booming demand for "structural steel beams" in a construction industry that only a few years ago still depended heavily on timber. A simple poll of steel beam wholesalers in the city of San José revealed that sales of these structural steel beams had increased by almost 41% from 1990 to 1993.

Many local timber wholesalers will in fact argue not that local prices have risen sharply, but that they will continue to lag behind unless "international economic prices" are used as a guide for pricing local timber. Some analysts assert that present imperfections in the timber market stand as a clear barrier, preventing local timber producers from receiving more realistic remuneration, and thus fostering unsustainable management practices in the tropical rain forests of Costa Rica (Stewart, 1994).

Major timber processors show no sign of abandoning their long-standing argument that barriers to the free local and international trade of logs have preserved the Costa Rican tropical rain forest and warded off a collapse of the local timber industry. Even if present policies continue, however, prices will in all likelihood continue to climb, and to the dismay of many, the rapid disappearance of local tropical rain forest acreage continues unabated.

For many years, the forest sector has been faced with numerous legal and "technical" arguments to prevent the free trade of many forms of forest products. However, in view of the rapid disappearance of tropical rain forests over the past 25 years, it is clear that cheap local wood has, in fact, promoted the destruction of the tropical rain forest. Many species that have now come into use were once wantonly destroyed and burned through the early use of highly selective logging practices. If they could have been saved from wasteful loss, probably the country would be in a better position today.

Several major problems make it very difficult to evaluate the performance of the forestry sector. First, there is very little statistical material, particularly information on local prices, for reasons that are well known but are not of interest for the purposes of this chapter. Second, timber is used for two major purposes: industry and charcoal manufacture. These two sectors carry different relative weights in determining the real average local price for timber products of the tropical rain forest.

The shortage of price information was approached in two ways. The total value of production for the forest sector in current Costa Rican Colones was obtained from the National Accounts Section of the Central Bank. Figures on total local wood production in cubic meters were compiled from the General Forestry Division of the Ministry of Natural Resources, Mines and Energy and FAO/Forest Statistics.

The gross production series was then divided by the production value series, giving the "local imputed price of total timber produced, per cubic meter." This was divided by the "average current exchange rate" reported by the Central Bank, to offset the effects of inflation. The imputed value per cubic meter was used as a proxy for the national average price of a cubic meter of timber for all uses. As can be seen in Table 5.3, the price level in dollar terms never rose above US$12 over the last decade, even though economic prices went as high as US$82 in 1990.

It is certainly true that such a wide gap between local prices and world prices provides no incentive to care for this poorly remunerated resource. Table 5.4 shows what would have happened to the total value of the forestry subsector's

Table 5.3   Financial and Economic Prices for Costa Rican Timber

Year	Economic prices (US$/m³)	Financial prices (US$/m³)
1980	89.20	14.90
1981	75.90	8.20
1982	63.10	6.40
1983	63.30	7.60
1984	53.90	9.90
1985	51.20	8.80
1986	55.60	9.80
1987	68.60	9.80
1988	72.20	9.20
1989	72.40	9.90
1990	81.70	10.80
1991	76.90	11.70

Note: Based on data from the World Bank and the Dirección General Forestal de Costa
     Rica.

Table 5.4   Forest Sector Participation on Sector GNP (in percent)

	Participation based on	
Year	Economic prices	Financial prices
1980	4.5	16.8
1981	3.2	19.5
1982	2.5	16.6
1983	2.6	12.6
1984	3.1	11.8
1985	3.1	12.2
1986	3.3	11.5
1987	3.0	15.6
1988	3.1	16.6
1989	3.1	15.3
1990	3.8	19.8
1991	5.4	24.5

Note: Based on data from the World Bank, the Dirección General Forestal de Costa
     Rica, and the Central Bank of Costa Rica. See text for explanation.

Table 5.5   **Forest Activity Nominal Protection Rates**

Year	Nominal protection rates
1980	−83.3
1981	−89.2
1982	−89.8
1983	−88.0
1984	−81.6
1985	−82.8
1986	−82.4
1987	−85.7
1988	−87.2
1989	−86.4
1990	−86.8
1991	−84.7

Note: Based on data from the World Bank, the Dirección General Forestal de Costa Rica, and the Central Bank of Costa Rica. See text for explanation.

output if the country's total production were valued at "world," shadow, or economic prices. As can be seen, at financial prices, the sector's share of total output climaxed in 1991 at a bare 5.8%, but at economic prices, the result the same year was an astonishing 24.5%. The result of such a situation is known as the nominal protection rate, or transfers from timber producers to the rest of society, in the form of prices. From 1980 to 1992, the figure ranged between −81 and −89 (Table 5.5). The negative sign denotes a transfer from society to timber producers.

Incidentally, these negative rates are also two or three times greater in magnitude than those of other products, particularly basic grains (Salazar et al., 1993). This clearly reveals that the forestry sector is at a great disadvantage compared with other agricultural subsectors of the economy during the last decade. The implications of all this are inescapable. A sector that has negligible macroeconomic clout receives very little attention in policy dialogues and in decision-making circles. This, coupled with the effect of low local financial prices and a notoriously imperfect market setting, will always prevent the sector from living up to its full economic potential. Sustainable forest management will remain elusive. If the sector were able to show its muscle in national accounts, it would have a far different degree of participation in the processes that shape its policy environment.

Because of its lightweight role in the economy, the forestry sector has rarely been invited to participate in policy arguments. One of the results of this is the environmental degradation that the country is now experiencing. Severely lop-sided development plans have inevitably emerged from decision-making processes that separate forestry from other policy debates, particularly agriculture, overlooking the environmental interplays that are well known to ecologists.

Notwithstanding these deficiencies, it is not enough merely to incorporate forests into the sectoral policy debate. A new, more holistic approach is needed. All the material cited above clearly reveals that the true value of timber resources has gone unrecognized during the last decade. The material coming out of ITTO resoundingly upholds the importance of a holistic approach and a more accurate assessment of true resource value (ITTO, 1994).

Any new policy dialogue will have to integrate all sectors of human economic activity if there is to be any hope of resolving our pressing environmental problems in coming years. In the future, the concepts of human and natural capital will have to be treated as one and dealt with accordingly.

Economic Prices and Cross-Sectoral Subsidies

Developing nations have discovered a number of "perverse" economic policies for transferring resources from one sector to another, and Costa Rica is certainly no exception, as the foregoing discussion shows. The negative protection rates in question were exacerbated by foreign exchange policies under which the currency was subjected to regular devaluations as a normal procedure to improve the country's short-term competitive position in agriculture and livestock and to promote the small number of industrial products targeted under the famous import substitution models of the 1960s and 1970s.

Under these devaluations, products with little export potential became even less attractive. Forest producers no longer had any incentive to care for tropical rain forests, and local timber production offered very few rewards. At the same time, macroeconomic policies were offering artificially high gains in exchange for destroying the rain forest and converting it to pasture or pressing it into use for other traditional tropical exports, such as African palm and cocoa.

Although available information is incomplete, there is enough data to perform preliminary calculations of transfers and subsidies from the forest subsector to the rest of the economy. The results can be seen in Table 5.6. These transfers and subsidies grew from C1.2 billion* in 1980 to C28.0 billion in 1991, or from C354 per cubic meter in 1980 to over C6000 per cubic meter in 1991.

Costa Rica is now engaged in a country-wide campaign to maintain its forest and provide a sustainable environment for future generations. The present government administration has put forward an innovative proposal to sell carbon

---

* C = colones.

Table 5.6 Level of Subsidies per Cubic Meter of Timber, in Constant 1990 Colones

Year	C/m^3
1980	30.5
1981	168.3
1982	443.7
1983	355.3
1984	356.3
1985	439.4
1986	593.3
1987	1360.5
1988	2254.4
1989	2397.5
1990	3121.2
1991	6103.3

Note: Based on data from the World Bank, the Dirección General Forestal de Costa Rica, and the Central Bank of Costa Rica.

emission rights to developed countries as a way to finance this important, worthwhile effort.

Many feel that while this proposal is undeniably important and interesting, it misses the point. It transfers responsibility outside, while failing to address the needs of Costa Rica's forests, forest dwellers, and producers. This sector has transferred much wealth over the past decade, receiving low prices for its timber and thereby subsidizing the rest of the national economy. The new proposals make no allowance for the rights of this sector.

If the government of Costa Rica is able to sell emission rights as proposed, the country's 1994 earnings will total approximately US$100 million or around C15.5 billion a year. This is equal to nearly half the 1992 subsidies. If society were to decide instead to pay the economic price of timber, a new era of sustainability for the Costa Rican forest could well be financed domestically.

In any event, a high price on wood for local use would probably deter present patterns of wastage and destruction and encourage better management of the country's forest. All parties would benefit. The argument that higher prices will simply accelerate the destruction of the forest appears ingenuous. In fact, rational economic agents do not plunder a valuable resource for short-term gain, when a program of sustainable forest management would shore up attractive long-term opportunities to earn income repeatedly from the same resource.

The argument that high prices trigger deforestation certainly is not borne out by farm surveys currently being taken. Instead, the prevailing argument appears to be entirely different. Even livestock producers whose farms preserve little or no remaining forest are thinking of abandoning the less productive areas, letting them revert to "nature" and be taken over by secondary forest growth. For the first time, they are eyeing the special treatment now given artificially reforested lands, asking that these same benefits be extended to secondary growth forest.

This trend has received even greater momentum as markets open for secondary forest species, which are now actually preferred in many cases, due to their rapid growth. In this new setting, many will be forced to rethink their attitudes toward the "forest," seeing its economic potential from a new economic and technical perspective.

Discussion

Economic Pricing and Valuation of the Macro-Forest

In the coming years, the move to price timber more appropriately will begin to converge with new activities based on extraction of other products from the tropical rain forest. This will give a whole new understanding of what an acre of forest is really worth and awaken new interest in planting trees.

A number of studies have already been published on assigning a more accurate value to the forest, so as to account for the macro and micro values of different forest products (Panayotou and Ashton, 1992; Godoy and Bawa, 1993). The World Bank (1993) recently completed a study in Costa Rica, combining local data with figures from other parts of the tropical world, and placed the real value of a hectare of tropical forest in the country at over US$2000.

CATIE is now reviewing these figures, based primarily on the results of its own local work (Aguirre, 1994). Although the figures differ somewhat from World Bank numbers, preliminary findings of both studies clearly indicate that the "total value" of an average hectare of tropical rain forest is far higher than it was considered in the past. In fact, in many cases, the forest was assigned a value of zero in order to justify the removal of forest cover in favor of other economic activities on land whose only ecologically sound use would have been forest.

Table 5.7 shows results obtained at CATIE, based on preliminary calculations, showing "flow values" rather than stock. This distinction is important to remember when evaluating the full dimension of the forest's real contribution to gross national product. Based on financial pricing, the total flow value of a hectare of tropical rain forest was estimated at around US$240 per year in constant 1990 dollars. Calculations based on economic pricing assign a social

Table 5.7 Macroeconomic Evaluation of a Hectare of Tropical Rain Forest in Costa Rica, in Constant US$1990 (percentages in parentheses)

Category	Value using financial prices	Value using economic prices
Carbon fixation	77.00 (32)	77.00 (9)
Wood production	39.28 (16)	625.10 (71)
Ecotourism	66.50 (28)	66.50 (8)
Electric energy	27.69 (12)	27.69 (3)
Water for domestic and rural use	27.80 (12)	27.80 (3)
Pharmaceuticals	2.31 (1)	2.31 (0.3)
Wild animals and plants	51.00 (21)	51.00 (6)
Total value	240.58 (100)	877.40 (100)

Source: Unpublished material from CATIE research.

value of about US$800, or 2.4 times more, to the same hectare. These figures compete very favorably with many of the non-forest uses currently being given to the land, particularly commercial beef production. Another important conclusion can be drawn from this analysis. If figures are based on economic prices, timber production as a share of the total value of the forest rises from 16 to 75%. Such figures make the idea of sustainable forest management even more appealing.

A 1992 study asked how much it might cost to preserve the country's biodiversity. The result? An estimated US$1 billion over a ten-year period, or just under US$100 million per year (INBio, 1992). We can use this figure as the total cost of preserving the remaining 1.3 million hectares in sustainable condition and express it in per-hectare terms as US$77 a year per hectare. This amount can be roughly covered by the revenue from selling carbon emission rights, if present attempts should materialize. The remaining benefits would remain as "ecological–economic" profit to society.

The real problem is that the benefits accrued to society are of little interest to individual timber producers and to the forest dwellers who, in the final analysis, are entrusted with keeping up and managing the tropical rain forest. Unless society is willing to make sure the social benefits are spread equitably, these people will have no incentive to practice sustainable forest management.

Implications of Forestry and Environmental Policy Design

Sustainable forest management can become a reality only if appropriate policies are identified and implemented. However, "hard economic facts" are necessary

CARL A. RUDISILL LIBRARY
LENOIR-RHYNE COLLEGE

if such political decisions are ever to be made. Society at large and politicians in particular need to be convinced that such policies are in the best interest of society, particularly in times of serious economic and ecological hardship. Four policy questions are outlined here that must be addressed if sustainable management of the forest is to become a central part of economic discourse:

1. *Price trends and resource allocations: timber vs. other activities*—Evidence already gathered clearly shows a future in which a number of economic forces will converge so that the sustainable production of wood for all uses will become an excellent business opportunity. For individual investors and society as a whole, private and public resources invested in the sustainable use and management of forests will be highly profitable in the medium and long term. If this is true, credit and investment policies for the forestry sector need to come under close scrutiny, taking into consideration the sector's diverse economic contributions. Particularly critical are policies that encourage both agricultural producers and urban dwellers to combine forest production with other activities. Each different activity needs to become an integral factor in the farm's cash flow, so that the final outcome will be attractive to timber producers and forest dwellers.

   Forest product marketing also needs attention, because of current expectations for improved price conditions. Markets need to be properly organized and structured so that the benefits of higher prices and more accurate value assessment will be distributed equitably and sustainably, particularly among small-scale owners and forest dwellers. Production and conservation need not be antagonistic.

2. *Raise the sector's profile in the policy dialogue*—If predictions of economic renaissance in the forest sector are borne out, the sector will need to play an active role in the general agricultural policy dialogue. This holds many implications, such as coordinating studies, engaging in the policy debate, and the participation of both public and private actors. Traditional policy design practices center the dialogue around agricultural activities. This is no longer environmentally sustainable. If the forest sector continues to be isolated from national and sectoral policy debate, conflicts over land and water use and the degradation of renewable resources will be inevitable.

   Policy forums for the agricultural sector will need to be all-inclusive, particularly as the forest becomes a major contributor to economic and environmental stability for Costa Rican society. Separate forums are neither practical nor acceptable over the long term, if sustainable policies are in fact a goal for the future.

3. *Comprehensive assessment of forest acreage*—If more accurate values are attributed to forest land, reflecting its full production potential, other issues will come into clearer focus:
   - *Sustainability of timber resources*—This would include measures and organizations to make sure that technical and economic measurements are taken into consideration, thus guaranteeing that all benefits are accounted for and equitably distributed.
   - *A careful balance of conservation and production*—All production processes will have to include conservation measures as part of their investment plans and consider these measures as valid medium- and long-term operating costs. Short-term lending policies will need to be reworked to guarantee that these activities are included in the regular portfolio of the banking system.
   - *Other non-timber activities*—The forest ecosystem is now recognized as a source of water, foreign exchange, pharmaceuticals, carbon fixing, and wild plants and animals for domestication and for educational and anthropological uses. These will all have to be counted as valid economic activities and private and public resources allocated for developing them.

4. *Appropriating social benefits to private holders*—Much of the benefit of the forest is of a social nature. However, it is the individual producers who, in the final analysis, must care for the resource, and they need to see that they have a share in the benefits. Mechanisms will need to be developed to create subsidies and introduce transfers from other sectors benefiting from the preservation and survival of the forest. Those who reap the rewards of the tropical ecosystem will need to understand this situation. The way things are, it is difficult to imagine that producers and dwellers of the forest will maintain favorable attitudes toward sustainability unless they see a direct short-term benefit for themselves aside from long-term benefits to society.

Agroforestry: A Contribution to Sustainable Development

The new appreciation of the value of trees, in the author's opinion, will extend beyond the forest. Over the past 40 years, CATIE has conducted extensive research on the role of agroforestry systems in developing sustainable agricultural systems, particularly for small farmers. Recent research has shown that trees can provide enough nutrients to small-scale basic grain production to save the country some US$38.80 per hectare. If trees were added to only half the area presently sown to corn and beans, the production process would benefit from billions of colones in nutrients obtained naturally (Dominique et al., 1994).

None of these promises will materialize, however, unless the value of trees is redefined, both as particular individuals or species and, in general, because of their capacity to generate wealth for the society of developing countries. Much can be achieved if ways are found for trees to make these and other kinds of contributions to the welfare of society. We need new ways of looking at trees and their contribution to human welfare.

Economic or Financial Pricing? Subsidies, Distortions, and Possible Solutions

The producers and dwellers of the forests have transferred substantial resources to other sectors of society. Such distortions will have to fade into the past if our new goals are ever to be met. First, timber marketing policies will have to be revised, and the perspective of sustainability will have to be maintained in an open market, without forgetting the concept of economic pricing and the benefits that such market policies can bring to all parties. Second, current local marketing mechanisms need to be reassessed to ensure competitiveness and equitable distribution of economic benefits. Third, possibilities need to be explored for replacing other products with forests, as prices move in favor of timber products and against traditional agricultural and livestock activities. Tradition will have to be reconsidered and secondary forest will have to be addressed as a research and policy issue, seen particularly as an appropriate land use in areas where primary forests should have never been removed in the first place.

Conclusions

The direct implications of the findings in this chapter are:

- It is highly unlikely that anyone will be motivated to care for or manage a resource that is as undervalued as the forest, whether for immediate use or future benefits, particularly in terms of an investment.
- There is a clear need to profoundly rethink trade and commerce policies for the Costa Rican forest sector.
- If economic pricing had been used in macrostatistics on the agricultural sector, the forest sector would almost certainly have had a much stronger voice in economic policy arguments that have affected the sector over the past five years.
- Easily adaptable economic mechanisms must be developed to reflect the macroeconomic and sectoral externalities of natural resources.

Macroeconomists will not readily translate a long, involved process with a strictly ecological orientation into the national accounting system.

- No assessment of externalities could have counteracted the initial miscalculations, based on what appears to be a real underpricing of timber resources.

- Economic pricing would have so improved economic returns for local timber producers that they would have refrained from clearing at least part of the land on their farms, particularly the areas suitable only for forest.

- A comprehensive policy dialogue must embrace the agricultural sector and address development issues of interest to foresters. Only then can the forest sector guarantee sustainable production and medium- and long-term contributions to the environment.

- The role of trees will have to be reassessed from a holistic standpoint if true ecosystem management concepts are to emerge, allowing production and conservation to operate in harmony.

Finally, it is important to understand that all other arguments about production and economic contributions of non-timber products from the forest are probably of "social interest" only. They contain little convincing information about "real" benefits for the owners of forest lands, on whose shoulders ultimately rests the sustainable use of the resource, with all its implications.

Real pricing means real money that can be acquired almost immediately. This is language that farmers, rich or poor and of any ethnic origin, can easily understand. "Real economic prices" will buy the time to develop sustainable technologies for real, long-term conservation of the forest.

References

Aguirre, J.A. (1993) Oferta, demanda y precios de la madera: el caso de Costa Rica. 1985/1992. Abstract, Semana Científica (Scientific Weekly), No. 19, CATIE, Turrialba, Costa Rica.

Aguirre, J.A. (1994) Metodología para la Valoración Macroeconómica/Ecológica del Bosque Tropical. CATIE Internal Working Document. MIRENEM/epc. Turrialba, Costa Rica, 110 pp.

CCT/WRI (1991) *La Depreciación de los Recursos Naturales en Costa Rica y su Relación con las Cuentas Nacionales.* CCT/World Resources Institute, San José, Costa Rica, 148 pp.

Dominique, J.R., Aguirre, J.A., and Kass, D. (1994) Evaluación ecológica económica de un sistema de cultivos en callejones asociando maiz (*Zea mays l.*) con Poró (*Erythrina poeppigiana* (walpers) O.F. Cook). *Agroforestería en las Américas,* in preparation.

Godoy, R.A. and Bawa, K.S. (1993) The economic value and sustainable harvest of plants and animals from the tropical forest: assumptions, hypotheses and methods. *Economic Botany,* 47: 215–219.

Godoy, R.A., Lubowski, R., and Markandaya, A. (1993) A method for the economic valuation of non-timber tropical forest products. *Economic Botany,* 47: 220–234.

INBio (1992) *Estudio Nacional de Biodiversidad. Costos, Beneficios y Necesidades de Financiamiento de la Conservación de la Diversidad Biológica en Costa Rica.* MIRENEM/MNCR/Instituto Nacional de Biodiversidad, Heredia, Costa Rica, 188 pp.

ITTO (1994) *Forest Resource Accounting: Stock Taking for Sustainable Forest Management.* World Conservation Monitoring Centre, Cambridge, U.K. and International Institute for Economic Development, Japan, 51 pp.

McNeely, J.A. (1988) *Economics and Biological Diversity: Developing and Using Economic Incentives to Conserve Biological Resources.* International Union for the Conservation of Nature, Gland, Switzerland, 234 pp.

Muul, I. (1994) *Tropical Forest, Integrated Conservation Strategies and the Concept of Critical Mass.* MAB Digest, No. 15. UNESCO, Paris, France, 82 pp.

Panayotou, T. (1993) *Green Markets. The Economics of Sustainable Development.* ISC Press, San Francisco, CA, 168 pp.

Panayotou, T. and Ashton, P. (1992) *Not by Timber Alone: Economics and Ecology for Sustaining Tropical Forests.* Island Press, Washington, D.C., 282 pp.

Salazar, M., Santana, C., and Aguirre, J.A. (1993) Protección a la Agricultura. Marco Conceptual y Metodología de Análisis Computadorizado. IICA. Programa I: Análisis y Planificación de la Política Agraria, San José, Costa Rica, 134 pp.

Stewart, R. (1994) *Incidencia del Comercio Internacional sobre la Economía del Sector Forestal Costarricense.* Stewart & Associates, Costa Rica, 118 pp.

World Bank (1993) Commodity Markets and the Developing Countries. Washington, D.C., March/June, 65 pp.

The Valuation and Pricing of Non-Timber Forest Products: Conceptual Issues and a Case Study from India*

Kanchan Chopra
Institute of Economic Growth, Delhi, India

Abstract

With preservation of biodiversity being recognized as an important objective of natural resource management, non-timber forest products (NTFPs) have come to occupy a significant position. From a welfare viewpoint, non-timber forest products have provided livelihoods to a large number of forest communities. The income and consumption available from NTFPs have complemented that from agricultural activity in the better forested regions of the country.

* This chapter is a revised version of a paper presented at the Third Biennial Meeting of the International Society for Ecological Economics, San José, Costa Rica, October 1994. The author wishes to thank the Department of Natural Resource Economics of the Indira Gandhi Agricultural University, Raipur for organizing the survey work in the villages of the region. Thanks also to the Tropical Forestry Research Institute, Jabalpur; the Centre for Minor Forest Products, Dehradun; and the office of the Ford Foundation, Delhi for access to unpublished material and data. The participants of a workshop organized by the Society for Wasteland Development in Bangalore also provided comments that helped in revision.

1-57444-077-2/97/$0.00+$.50
© 1997 by CRC Press LLC

This study focuses on the valuation and pricing of NTFPs, with particular reference to Raipur, a forest-dominated region in central India. After discussing alternate concepts of value, it is postulated that, in general, user valuation of NTFPs is appropriate in situations where they are treated as renewable resources. Markets are found to be a significant influence on users of NTFPs in Raipur because a large proportion of forest produce finds its way to market. From the viewpoint of efficiency, the best welfare intervention is found to be enabling the markets to function smoothly. Intertemporal markets do not perform so well: they may lead to overextraction in the present. Methods of estimating option value are suggested, and, further, by evaluating forest investment projects in Raipur in the context of alternative forms of management, it is found that even if forests were managed for NTFPs alone, investment in their regeneration would be economically efficient. Trade-offs between timber and non-timber products are examined, and, finally, implications of introducing biodiversity valuation into policy objectives are discussed.

Introduction

Forests are traditionally considered as the suppliers of timber. Perhaps due to the overwhelming significance of the market as an economic institution, this function of forests has been given somewhat exaggerated importance. Of late, however, a number of other goods and services have come under scrutiny. The services of nutrient recycling, management and preservation of water cycles, regulation of microclimates, and the production of non-timber forest products have been highlighted. Additionally, as forms of monoculture are becoming more and more prevalent on cultivated land, forests are the major natural repository of large components of the remaining gene pool of plants. Consequently, with the preservation of biodiversity being recognized as an important objective of natural resource management, non-timber forest products (NTFPs) have come to occupy a significant position. All kinds of tropical forests occupy a critical niche in this context because tropical forests are distinguished by the availability of a large variety of NTFPs. India, for instance, possesses about 320 of the world's 425 families of flowering plants, or about 21,000 species, of which 3000 are known to yield NTFPs.*

From a welfare viewpoint, NTFPs have provided livelihoods to a large number of forest communities. The income and consumption available from them

---

* See Gupta (1991) for details of the non-wood forest products in India. See also Prasad and Bhatnagar (1991a). Details of economic/commercial NTFPs and medicinal NTFPs are given in the appendix to this chapter.

have complemented that from agricultural activity in the better forested regions of the country. Forest communities have derived sustenance from NTFPs in periods of stress and have used NTFPs as inputs or raw materials into the production of items of daily use in normal times. In effect, NTFPs, if harvested and used judiciously, are a kind of renewable resource available for exploitation from year to year. These products, therefore, have to be valued from a number of different viewpoints, each giving rise to a set of significant analytical and empirical issues.

This study examines issues related to the pricing and valuation of NTFPs, with particular reference to the forest-dominated region of Raipur in central India. The analyses herein draw on a combination of field data collected in Raipur since 1990 and data from other studies. It is shown that some concepts of value justify a focus on the manner in which local communities view forest resources, while others require a national or even global approach.

The chapter is divided into six major sections. First, the nature of NTFPs is examined, and different concepts of value relevant to them are discussed. The concept of *user valuation* is introduced. Second, the kinds of NTFPs found in Raipur are described. The third section analyzes the applicability of user valuation to NTFPs and discusses the efficiency of the market channels for NTFPs in and around Raipur. Fourth, methodologies for the determination of value with imperfect intertemporal markets are examined, and the problem of overextraction in the present is discussed. Fifth, the links between forest management, supply, and evaluation of NTFPs are investigated, with particular reference to the sal forests in Raipur. Trade-offs between timber and non-timber products are examined, and it is found that even if the only benefits accruing from these forests were NTFPs, their regeneration would still be a profitable investment.

Finally, some implications of introducing biodiversity valuation as a policy objective are discussed, in particular, the complex trade-offs at the national level and the potential for evolving strategies at the international level.

Concepts of Value: Relevance in the Context of NTFPs

Various concepts of value for different kinds of natural resources are described in the literature. For example, instrumental value emerges from use in the present and/or the future. Non-use value, on the other hand, is related to the utility emerging from pure existence.* The nature of the resource is of the essence in

---

* These are components of total economic value, which consists of use value, non-use value, and option value. For details and discussions, see Pearce and Turner (1990) and the earlier works of Arrow and Fisher (1974) and Bishop (1982).

determining the relationship between the types of value. Substitutability between present and future use arises either when a resource is exhaustible or when it is renewable but acquires the character of exhaustibility due to overextraction. If the extraction rate is not too high, renewable resources permit a relatively constant rate of extraction over time.

NTFPs are usually renewable resources. They could, however, acquire the characteristics of exhaustible resources if the rate of extraction is higher than that which is sustainable, the latter being defined as an ecologically determined rate of extraction that does not harm the future development of the forest. In the case of some NTFPs, it is not often possible to extract more than what is produced in a certain year; examples are leaf- and flower-based products. This ensures sustainable use. In a few other products, such as lac- or resin-based ones, overextraction is a more real possibility, and the issue of substitutability between value in the future and in the present becomes relevant. It is in this situation that option value and bequest value are relevant.

Further, while the sustainable rate of extraction is ecologically determined, actual rates of extraction depend on economic and institutional variables. Empirical evidence on the issue of sustainable use or otherwise of NTFPs from across the Latin American, African, and Asian forests seems to yield diverse kinds of results. While a number of anthropological studies support the viewpoint that forest communities do not overharvest NTFPs, some recent evidence also suggests that they may indeed do so in some situations.*

The trends in extraction of NTFPs over a period of time depend on factors related to both demand and supply. On the demand side, the per-capita demand depends on income levels. Consumption of many plants and animals falls when income increases because people substitute forest goods with cheaper substitutes. On the supply side, the opportunity cost of labor spent in collection increases with the availability of other kinds of work. The increase in income, then, could be positively or negatively associated with the level of extraction.

It is argued in this chapter that the concept of instrumental value relevant for NTFPs from the viewpoint of society depends on the level of extraction. As long as extraction is at less than sustainable levels, use in the present is preferred to use in the future. Consequently, consumption value is at a premium. This can be approximated either by use value to the consumer, as in a non-monetized market, or by exchange value if monetized or barter exchange exists. Like any other commodity being put to present use, an NTFP possesses a use value and an exchange value. The use value depends on the function that it performs, rather than the psychic value, or utility, of that function for the user. It is an instrumen-

---

* See the studies listed in the special issue of *Economic Botany* (1993), for example, those by Godoy and Bawa with respect to Latin American countries.

tal value of the subjective kind. The exchange value depends, in addition, on the conditions of exchange; in other words, the nature of the market in which a product is traded. The viewpoint adopted in this chapter is that the relevant value is user valuation. The concept of value that is most relevant is that which the user considers to be most significant in his situation. If the user sells a part of his product, the market or exchange value is of relevance; if he collects it for his own consumption, the use value should be the focused. In other words, it is the revealed preference of the user that is given primacy.*

Once overextraction becomes a reality, there is a cost attached to use in the present. Put another way, there is a benefit to be derived from preservation. The individual does not take account of this in his reckoning of use, but society must do so. This constitutes a source of deviation between social and individual values. This deviation is in effect an expression of the inadequacy of the market to reflect appreciation in value in the future compared to the present.** Social valuation then necessitates the determination of an "option value."

The existence value of NTFPs is a non-instrumental value arising out of the amenity value of forests or out of their contribution to biodiversity. It is possible to maintain that instrumental and non-instrumental values are positively related; that is, to treat the existence value as a positive externality, but this is not always so. It depends on the management system of the forest and the relative magnitude of timber and non-timber products that a forest is capable of yielding under alternate systems. This is an exogenously determined condition in the case of natural forests. In the case of managed ones, however, silvicultural practices being followed determine this ratio. And in a situation where the market dominates, silvicultural practices established over a period of time tend to be biased in favor of timber. This situation is self-perpetuating, which introduces a divergence between the present exchange value of timber and the existence value of a mixture of timber and non-timber products. A second major correction to user valuation needs to be made, therefore, in the context of the optimal management system.

The appendix to this chapter gives the function or the use of NTFPs in the Raipur region of India perceived to be most significant by different users and the correspondingly significant concepts of value. It should be pointed out that in the case of multiple uses, the "dominant" one has been listed. As long as other

---

* Such an approach is being considered by recent and ongoing studies, in particular in developing economies; see, for example, Reddy (1994).

** One way to account for this imperfection in the market is to use lower rates of discount, or to dispense with discounting altogether, as often stated in the literature on natural resources. The author feels, however, that this can result in distortions in other investment decisions as well. Allowing for a cost of present use is a better method of adjustment.

ancillary considerations do not conflict with this one, it is always possible to accommodate them within the framework of the dominant one.

This chapter initially focuses on user value to the forest community. Later, deviations between social and market value arising from the special characteristics of NTFPs as "semi-renewable" resources are examined. Two aspects that are discussed at length in the context of Raipur are (1) the option value of certain kinds of NTFPs and (2) the links between alternative systems of forest management and supply and therefore the existence value of NTFPs.

The Region and Its Important NTFPs

Raipur is situated in the state of Madhya Pradesh in central India. As listed in Table 6.1, it has a geographical area of about 2.2 million ha with red, sandy soils and an average annual rainfall of 1100 mm. Agriculture and allied activities constitute about 80% of total economic activity in the district. With a population of nearly four million, 80% of which is rural, the forest base, and its implication for rural livelihoods, is of utmost significance in this district. Forest-based activity complements agriculture as a major source of income, more so as the tribal population constitutes around 18% of the rural population. As a proportion of the geographical area, forests comprise 32%, three-quarters of which is characterized as dense forest and the rest as open forest.

The forests of Raipur belong to the so-called "sal" region in Madhya Pradesh. Sal, with its associated species, dominates the forests in this region. Bamboo (*Bambusa arundicea*) is found mostly as an understory. There exists a large

Table 6.1   Raipur District, General Indicators

Geographical area	22,158 km^2
Forest area	6,859 km^2 (32% of total)
Net sown area	9,391 km^2 (42% of total)
Population	
Urban	771,620 (20% of total)
Rural	3,136,420 (80% of total)
Total	3,908,040
Working population	1,842,400 (47%)
Mean population density	183 people/km^2
Annual population growth rate	2.4%
Literacy rate	48%

Source: Centre for Monitoring the Indian Economy (1993).

variety of climbers, grasses, and other undergrowth, creating an environment for a large array of non-timber products. Different parts of the forest result in the production of a number of different kinds of NTFPs: the leaves, the fruits, the roots, the understory, the bark, and the shoots each give rise to products that can be used both as intermediate- and final-use goods. Further, the existence of this array of products increases the biodiversity value of the forest. Details of a cross-section of products, representing those used by rural communities in forging livelihoods and those representing market and industrial requirements, are listed in the appendix.

While data on the production of NTFPs for Raipur district could be put together only for one point in time, time trends in the production of some important products could be studied at the state level. Three such commodities are tendu leaves (leaves of *Diospyros melanoxylon*), bamboo, and sal seed (seeds of *Shorea robusta*). Regressions were fitted to data for 17 to 26 years* for the production of these three commodities. Tendu leaf collection showed an upward trend, whereas bamboo did not display any trend. Sal seed collection fluctuated with a cycle of two to three years. However, if the 1970s and 1980s are viewed separately, it is found that production stabilized in the 1980s even as the resources were declining. Sustainability therefore has to be viewed as a commodity-specific phenomenon, and time series of production must be available to arrive at any reliable results. In the absence of such information at the district level, it is difficult to hazard a guess.

With a view toward illustrating the issues involved in the valuation of NTFPs, a few important ones have been selected for more extensive study. These belong to both the intermediate-use and final-use categories of products. Important among the selected intermediate-use products are tendu leaves (used mainly for the manufacture of an indigenous cigarette), sal seeds (which yield oil), harra (a tree-bark-based tanning material), and gums and resins. These are also the four nationalized NTFPs in Madhya Pradesh. Other intermediate goods analyzed are bamboo, sabai grass (*Eulopsis binata,* used for rope making), other grasses, and the leaves of *Bauhinia vahilii* (the mahul) used for making leaf plates and cups. Some products, such as bamboo, are demanded both by the larger organized sector (e.g., paper and rayon mills) and by small cottage industries. Natural forests supplement human food supplies, too. Products significant in this context in Raipur are *Emblica officinalis* (locally called aonla), *Buchanania lanzan* (known as chironji locally), and honey. The flowers of *Madhuca indica* and *M. longifolia* trees (locally called mahua) are another important source of sustenance in rural areas, in particular in times of stress (e.g., droughts). Table 6.2 gives the ap-

---

* The series was obtained from Prasad and Bhatnagar (1991a).

proximate production of these NTFPs selected for analysis, valued at Madhya Pradesh state prices.*

NTFPs as Renewables: User Valuation, Markets, and Other Institutions

Treating NTFPS as renewables, user valuation is estimated by attempting to find out how the collector views an NTFP. Such a user's perception of the value of the product may be in terms of either use or exchange. As communities become more and more monetized, and as linkages with markets are forged, the distinction between communities' perceptions of use value and market value is reduced. Sale becomes an available option through which purchasing power is obtained, which in turn can be used to acquire access to other necessary articles of consumption. It is, in the final analysis, the perception of the value of a commodity, whether for self-use or exchange, that should be paramount in determining the utility that accrues from it.

The value of NTFPs, whether it be use value or exchange value, accrues first and foremost to forest communities, although they are also consumed by larger units both within the district and outside.** As long as factor and product market imperfections continue to exist, this value will continue to be person and product specific.*** We shall therefore, in the first instance, attempt to evaluate these products (both as consumer goods and intermediate goods) from the viewpoint of forest communities.

Data Sources and Products

The user valuation part of the study depends on the following sources of data:

1. A survey conducted in two villages in the Nagri *tehsil* of Raipur district in 1990–91.†

---

* See the appendix for a more exhaustive list. Prasad and Pandey (1987) give details of medicinal plants, and Prasad and Bhatnagar (1991b) list wild edible products. Further, the concentration of plants of different species is studied in Prasad and Pandey (1992) to examine the biodiversity aspect of NTFPs. See also Biswas (undated).

** Some attempts at valuation of non-timber forest products at a macro level have been made. See, for example, Das (1992) and Chopra (1993).

*** An analysis of such imperfections in trade between Nepal and India is found in Edwards (1993).

† See Marothia and Gauraha (1992) for details with regard to the survey. The villages where the study was conducted (Nirrabeda and Farsiya) contain mostly tribal populations.

Table 6.2 Production of Important Non-Timber Forest Products in Raipur

Product	Units (in thousands)	Production	Sale price (Rs/qtl.)[a]	Value (Rsm)
Nationalized products				
Gum	qtl.	b	2000	—c
Harra	qtl.	a	140	—
Sal seed	qtl.	127	200	123
Tendu leaves	std. bag	304	2500d	3696d
Non-nationalized products				
Aonla	qtl.	1.46	700	5
Bamboo	tons	—	—	143e
Broom	—	—	—	2.7e
Chironji	qtl.	0.06	7000	2
Grasses for rope	—	—	—	0.56e
Honey	qtl.	0.08	4000	1.6
Karanji	—	—	1950	—
Kusum seed	qtl.	1.28	1150	7
Lac	qtl.	1.78	6200	54
Mahua flower	—	—	550	—
Mahul leaves	qtl.	6.1	1200	36
Palas seed	qtl.	0.21	500	0.52
Tamarind	—	—	725	—

Note: Financial data are given in rupees (Rs) where R1 ≈ $0.03 (1995 exchange rate). Rsm is millions of rupees. Data are estimated from secondary sources, in particular Gupta (1991), Prasad and Bhatnagar (1991a,b), and Madhya Pradesh State Minor Forest Products Trading and Development Corporation (1991).

[a] Production in the state was 126,000 quintals in 1986–87, valued at about Rs37 million, or Rs295 per quintal.

[b] Estimates for the items are computed as average expenditure on inputs by forest-based cottage industries (FBCIs) multiplied by the number of FBCIs. Production in the state is around 800,000 tons.

[c] Data not available.

[d] Sale price of tendu is in Rs per standard bag.

[e] Price may vary with quality from Rs1900 to Rs2500 per standard bag.

2. A resurvey conducted in selected households of four villages in the Gariaband range in West Raipur in 1994.*
3. Secondary sources of data from different studies for validation.**

Information was collected for a large number of NTFPs. For the in-depth study, the following three sets of representative products were selected:

1. The four nationalized products—tendu leaves, sal seeds, harra, and gum
2. Non-nationalized consumer good— aonla, tamarind, mahua flower, karanj seed, kusum seed, palas seed, honey, and chironji
3. Intermediate products—mahul leaves, lac, bamboo, and grasses

A look at Table 6.3 makes it clear that one product or the other is being collected and either marketed or used for self-consumption throughout the year. In classifying the NTFPs, tendu leaves could be placed in a separate category since almost the entire population is involved in harvesting them during the collection period of about a month. Additionally, a large number of rural families, in particular those of agricultural laborers, rural artisans, and marginal and small farmers, are NTFP collectors. More than 98% of those interviewed mentioned collection of NTFPs as a secondary source of livelihood.

Household income from collection of NTFPs came to 40% of total income on an average, the range varying from 11 to 53%, even when activities such as tendu collection were excluded. This average is obtained as a weighted average of income from different commodities, the weights being probabilities of households collecting particular commodities.*** All available evidence seems to indicate that collection of NTFPs certainly has the status of a secondary means of livelihood, even in one of the relatively affluent villages.† As shown in Table 6.4, the quantity traded as a percentage of that collected varies from 35 to 100%.

All available evidence therefore indicates that NTFP collection is not only part of a survival strategy followed in periods of stress but constitutes a legiti-

---

* These villages are a part of Bindraanavagrah and Darripara panchayats. The villages are Navagarh, Khamaripara, Darripara, and Kosami. Fifty households were covered in this resurvey. These villages are closer to a market town than those covered in the earlier survey. The villages selected during the 1990–91 survey could not be sampled because of law-and-order problems.

** In particular, the ORG study on forest-based cottage industries in Madhya Pradesh was of great use, as well as publications from the Centre for Minor Forest Products and Gupta (1991) (see references).

*** See Table 6.4 and the percentage of households collecting particular commodities given in column 4 of that table. Estimated from survey data.

† Indirect evidence was used to characterize a village as "affluent" (e.g., residents in a village did not remember the year of the last drought and/or one in which they had the purchasing power to obtain goods from local shops).

Table 6.3   Collection Calendar for NTFPs from a Sampled Household in Raipur, 1993–94

Product	Collection months
Nationalized	
Gum	June
Harra	February 15 to March 15
Sal seed	Mid-May to second week of June
Tendu leaf	April 15 to May 30
Non-nationalized	
Aonla	September to October
Bamboo	June to July
Chironji	March
Grass for rope	September to October
Honey	March
Kusum seed	June to July
Karanj	July
Lac	February to March; May to June
Mahua flowers	March 15 to April 15
Mahul leaves	Mid-April to mid-June

Table 6.4   Household Income from Collection of NTFPs in Raipur

Commodity	Mean income per household (Rs/annum)	Mean income per household as percentage of average rural income	Coefficient of variance in income among sampled households	Percentage of households sampled that collect commodity	Quantity traded as percentage of quantity collected
Aonla	724	14	134	60	86
Chironji	2,678	54	120	48	100
Harra	1,730	35	441	72	—[a]
Honey	5,269	106	137	80	—
Kusum	550	11	104	56	—
Lac	71,062	—	58	12	88
Mahua flower	2,675	54	147	98	16
Sal seed	1,011	20	139	96	—
Tendu leaf	29,001	581	117	100	—

[a] Data not available.

Sources: Columns 1 to 4, field data from January 1994 (see text); column 5, Marothia (1992).

mate part of economic activity in normal times as well. With this empirical evidence, it would be appropriate to examine the price that collectors are able to obtain through different market channels as an indicator of the purchasing power generated by the labor spent in collection. User valuation of these products has two components: a consumption value of the part used in self-consumption and an exchange value for the marketed part. NTFP collectors are also aware of the exchange value of the goods they collect. As in any other market economy, this purchasing power can be used to acquire access to any commodity. In such a situation, it seems appropriate to consider the market price received by the collector as an index of the value of the product to the seller. Since the efficiency of alternative market channels determines this to a large extent, it is examined.

The price received by the collector under different market channels is identified as an index of the efficiency of the corresponding channel. The alternative channels could be:

1. Sale to a tribal agent, who sells to a consumer*
2. Sale to an agent of a wholesaler
3. Direct sale to a primary retailer in a local market
4. Sale to a primary wholesaler, who sells to a secondary wholesaler
5. Sale to a secondary wholesaler

Value of the collected part of NTFPs considered relevant in the local market context is then computed under these alternative sets of prices. This is the value that the market assigns to the collected part of the product that accrues to the local people. Regarding non-nationalized products, five representative products are taken from the 1990–91 survey. The study identifies the market channels for each of these. Table 6.5 gives the market channels relevant for each of the five and the price obtained in each of the channels. The last column gives the collector's income as a percentage of the retail sale price. This is an average of the prices prevalent in Raipur market.

It is found that:

*   The retailer's margin as a percentage of the sale price varies from about 8 to 25%.
*   The collector's price as a percentage of the final sale price varies from 26% to as much as 85%.

---

* Studies such as those of Biswas (undated) and Chakravarty and Prasad (1989) describe the link between the tribal economy and the collection of minor forest products.

** Rs denotes rupees.

Table 6.5   Marketing Channels for Important NTFPs

Product	Channels	Collector's income per unit (rupees)	Collector's income as % of retail price in Raipur
Aonla (qtl.)	1	186	27
	2	186	27
	3	260	37
	4	214	31
	5	258	37
Chironji (qtl.)	5	6000	86
Mahua flower (qtl.)	3	300	50
Lac (qtl.)	2	5800	—[a]
	4	6000	—
	5	6200	—
Sal seed (qtl.)	6	—	
Tamarind (qtl.)	1	600	86
	3	700	100
	4	600	86
Tendu leaf (std. bag)	6	250	10

Market channels
1. Collector $\longrightarrow$ tribal agent $\longrightarrow$ primary wholesaler/retailer $\longrightarrow$ consumer
2. Collector $\longrightarrow$ tribal agent $\longrightarrow$ secondary wholesaler/commission agent
3. Collector $\longrightarrow$ primary wholesaler/retailer $\longrightarrow$ consumer
4. Collector $\longrightarrow$ primary wholesaler/retailer $\longrightarrow$ secondary wholesaler/commission agent
5. Collector $\longrightarrow$ secondary wholesaler/commission agent
6. Nationalized
7. Purchased from Forest Department

[a] Data not available.

Sources: Field data from January 1994, Marothia and Gauraha (1992), and secondary sources.

- The chain of intermediaries between the retailer and the collector varies considerably from product to product and seems to be smaller in the case of higher value products.

Note, for instance, that in the case of chironji, which sells in the retail market for Rs6,000** (US$200) per quintal, the collector's income is as high as 86% of this sale price. The same is true of lac, which has just two alternate market channels. Aonla, on the other hand, which has a number of channels and which also happens to be lower priced than the other two (Rs700 or US$23) per quintal, is able to fetch the collector only 27 to 39% of that price. Mahua flower also has a collector's price of about 50% of the retail price.

It can be argued that where the collector is aware of the high value of the commodity, he puts in the extra effort required to obtain access to the information with regard to price. An analysis of the retailer's margin also corroborates the above statement. Retailer's margins seem to be lower in commodities that have a higher price. Gum, lac, karanj seed, and honey are products other than those mentioned above for which the retailer's margin is lower. There could, of course, also be other reasons for the difference in the levels of imperfection of the different markets, and this needs more in-depth study.

It must be added that a higher price does not necessarily imply a larger total contribution to income. In the case of a number of the high-value items, annual collection is small and limited to a few days in the year.

Nationalization of NTFPs as an Intervention in the Market: The Case of Tendu Leaf

Tendu leaf is one of the four major products whose collection was nationalized in an attempt to protect the interests of collectors.* The price at which the leaf is sold to so-called "Bidi" cigarette factories varies with the quality of the leaf and the market conditions. It is found that labor income as a proportion of this price is not very high: it is in the range of 10 to 13%. The average collector gets Rs250 per standard bag. The collector may be better off now to the extent that this price is higher than it was in the pre-nationalized situation, and he is protected from year-to-year variations in market price. However, for products where other kinds of market channels exist, his income as a proportion of retail price is higher as a consequence of an increased integration of markets and a freer flow of information. Governmentally run institutions also increase overhead costs. On balance, a better way of looking after collectors' interests is to create conditions for more efficiently functioning markets.

The limited role of markets is to ensure that collections reach the consumer with efficiency and the collector gets an appropriate share in the final price. It can be added here that while market channels throw light on the efficiency with which collected produce is brought to the market, there is little they can reveal with respect to the sustainability of the amount collected. Whether or not the

---

* The Madhya Pradesh State Minor Forest Products Corporation organizes the collection of these products from cooperatives. The Forest Department is involved heavily in the collection of tendu leaf.

Table 6.6 NTFPs as Inputs into Raipur's Forest-Based Cottage Industry (FBCI)

Product	Average value (Rs)	Average expenditure	Value added per FBCI (Rs)	Total FBCIs	Percent income from FBCIs
Bamboo	6,230	1,246	4,984	23,668	54
Honey	107	32	75	11,541	3
Rope making	1,250	20	1,230	5,773	6
Grass works	4,440	178	4,262	not avail.	29
Broom making	2,446	107	2,339	5,195	40

Note: Economic data given in rupees (Rs).

Source: Operation Research Group (1993).

collection can be sustained from year to year is a matter to which markets, with a focus on the present, do not give any significance. That is a matter for social intervention, through institutions of another kind, either state or of a cooperative nature.

Providing Concessional Access to NTFPs as an Intervention in the Market

The user value of intermediate goods is equal to value added to production processes as a consequence of their availability. This is a measure of the welfare or income generated by such goods, given the state of labor markets and levels of information and development.

A large number of cottage industries depend on forest products. Four important instances of such industries in Raipur are bamboo-based products, honey collection, rope making, and grass-based products. An Operation Research Group (ORG) survey (1993) reported that about 23,700, 11,500, 5800, and 5200 units of these four categories existed in the district. They contributed on average 54, 3, 6, and 29% of the income of the households, respectively. Table 6.6 gives the details with regard to the operation of these units. Value added per forest-based cottage industry is given.

At times, intervention in the form of non-market institutions may be in place, ostensibly to increase value accruing to the collectors and to "protect" their interest, in particular when it is perceived that the exchange value is higher than the value that accrues to them. For instance, a system of distribution of forest-based products to village people who had traditional access to them (called "nistar") is in place in Madhya Pradesh.* This channel of provision is not a market in the conventional sense of the term. It is a system of providing free or concessional access to permanent residents of a certain area. The system, however, does not now function as it was intended to. A number of factors have led to both an increase in the price and a decrease in the supplies of most of the commodities that are available to the people. The case of bamboo is of special interest since it is in demand by the large paper mills in the state as well. The Forest Department has long-running agreements for committed supply to these units. The consequence, in situations of non-increasing supply, is that only a small part of the demand by forest communities for bamboo can be met from nistar supply. The price has also been increasing over time. It is now close to the average price realized by the department from the combined sales from industrial and commercial bamboo.** The ORG study referred to above reports, for instance, that only about 53% of the bamboo requirements of forest-based cottage industries are met by supplies from the Forest Department. Self-collection and purchase contribute to the rest.

It is clear, therefore, that as the market institution is strengthened, it tends to replace and weaken any other institutions that may have been in place. From the viewpoint of evaluation, this situation has the following implications:

- As integration with the market increases over time, earlier institutional arrangements, whereby "preferred" communities were likely to derive a higher user value than the price, are no longer feasible.
- As a consequence, it is appropriate to take the market price of products as relevant.
- The reduction of market imperfections seems the only route available for ensuring that a larger part of the exchange value accrues to the collector, both as consumer and as user of NTFPs as intermediate products.

NTFPs as Exhaustibles: Present vs. Future Use

NTFPs under judicious extraction are viewed as renewable. As discussed above, however, overextraction can take place in a number of situations. It can happen

---

* See National Centre for Human Settlement and Environment (1987). The rights of certain communities to occupational nistar is recognized, and bamboo is available at a concessional price to them through forest depots.

** See the reports of consultants in the Madhya Pradesh Integrated Forestry Project (1993), in particular those of Singh, Khare, and Chopra therein. Khare makes it clear that "a shift towards changing market prices from nistari right holders has already begun." This is perhaps inevitable, given the high demand from all segments and the compulsions of political economy.

in general whenever the signals given by the existing institutional structure emphasize the present at the cost of the future. This tendency may be strengthened whenever NTFP collectors belong to low-income groups and have few other opportunities of livelihood generation. A typical situation is one in which traditional norms of extraction have been violated due to the all-pervading impact of the market. While it is sometimes argued that development may have an overall positive impact on rates of extraction,* overextraction seems to be a distinct possibility.

This occurs when the developed markets are those in presently available goods and services; intergenerational preferences are not expressed in the market or, where expressed, seem to favor the present. Such a situation leads to overextraction in the present, as has happened in the case of some products in India and in particular in Madhya Pradesh. Once this happens, NTFPs are to be treated as exhaustible products. Consequently, use in the present is at the cost of use in the future. The notion of "user cost" becomes relevant.

Use in the future is an extension of the concept of instrumental value over time. If preservation is interpreted as future use and development is conceived of as present use, the valuation of a resource conceived as exhaustible must have a development value and a preservation value. This preservation value can be expressed in two ways: as the ratio of two alternative time streams of extraction and as the annualized preservation value, to be compared to the development value. In both cases, the preservation value expresses a value judgment in favor of intergenerational equity. We shall look at both alternatives.

Two income streams can accrue out of a resource: one over an infinite period of time and the other over a limited number of years. Suppose the annual value of the first is R_1 and that of the second is R_2; then R_1 is usually less than R_2 in particular if the resource is exhaustible. Further,

$$\text{NPV}_1 = \frac{R_1}{r}$$

and

$$\text{NPV}_2 = \frac{R_2 (1 - e^{rT})}{r}$$

where NPV_1 and NPV_2 are the two net present values, T is the number of years in which the resource would be exhausted if used at rate R_2, and r is the social rate of discount. The ratio of NPV_2 to NPV_1 is the opportunity cost of cutting present use to sustainable levels. With the ratio of R_1 to R_2 known, this can be found. Table 6.7a gives this opportunity cost with $R_1/R_2 = 0.8$. It can be seen that as the value of R_1/R_2 decreases, this preservation value increases, implying

---

* See the work of Godoy (1993) and others on Latin America.

Table 6.7 Hypothetical Examples of Relative Opportunity Costs and Preservation Values

r	T = 10	T = 20	T = 30
(a) Hypothetical relative opportunity cost of present rates of extraction			
12	0.45	0.67	0.75
10	0.5	0.69	0.76
5	0.66	0.75	0.78
(b) Hypothetical relative preservation value of overextracted NTFPs			
12	1.43	1.1	1.03
10	1.58	1.16	1.05
5	2.54	1.58	1.29

Note: See text for explanation. r is the social rate of discount in percent; T is the number of years in which resource is assumed to become exhausted.

thereby that the larger the difference between rates of sustainable use over an infinite period of time and the rate of use within a time T, leading to exhaustion, the higher the value of preservation. As the social rate of discount decreases, the ratio of the preservation value to the present use value (NPV_2 to NPV_1) increases; at low rates of discount, the future is preferred to the present.

The same trade-off between present and future use can be expressed in another manner, if non-use of the resource for a certain period is expected to result in sustainable production beyond that point. This happens when overextraction takes place and regeneration can be expected only by abstaining from extraction for a time length T.

The annualized development and preservation value in the case of use over a period of T years is $r/(1 - e^{-rT})$. If the resource were used over an infinite time period, the value would be r. Using this as the numerator, the preservation value becomes $1/(1 - e^{-rT})$. Table 6.7b gives the sensitivity of this preservation value (analyzed as the ratio of incremental production made possible by non-extraction) to rates of discount and the time period T.

The ratios given in Table 6.7b have significant implications for policy relating to the valuation of NTFPs in the phase of overextraction. With a social rate of discount of 12% and a time horizon of ten years, the value of use at the end of ten years would be equal to that now if production were to increase to about 1.4 times the present rate. By abstaining from present use, the natural process of regeneration is given a chance to operate and additional value is generated in the future. It is as if the decrease in present consumption is a kind of investment in the future, comparable to investment in capital goods. It is, in other words, an investment in natural capital. The above line of argument assumes, of course, that the capability to regenerate exists in the natural environment or that degradation takes place within certain limits. At times, additional investment in capital of a complementary nature may be essential for such a process to be initiated.

Several NTFPs in Madhya Pradesh are known to be subject to overextraction, among them gum obtained from tapping of the tree species *Sterculia urens.** The tapping of this tree for gum was banned in 1982 for a period of ten years. It is of some interest to see what the implication of such a policy is for the implicit value of preservation if the social rate of discount is taken to be 12%. With a time horizon of ten years, the implicit preservation value is 1.43. If, on the other hand, it is known from independent evidence that the production was expected to double after this ban of ten years,** it can be concluded that the social rate of discount implicit in the policy decision of imposing a ten-year ban on production is in the range of 7%.

Another NTFP supposed to be subject to overextraction is bamboo. There is evidence of gregarious flowering and decreased regeneration resulting from overextraction in many parts of the state. The value of preservation relative to present use would constitute an even more interesting case here as the latter has a distributional component as well. It is demanded by large organized industries as well as by the forest-based cottage industry. About 35% of total production goes to the paper and rayon industries, 20% or so is used in rural areas, and another 8% is used as fuel by relatively lower income households. Each of these categories of present use must be given a different weight, both relative to each other and relative to use in the future.

NTFP Valuation and Forest Management: The Role of Supply

As discussed previously, the valuation of NTFPs depends on whether they are to be viewed as renewables or exhaustibles. While this depends largely on permissible rates of extraction determined by biological factors, it is also affected in a very crucial manner by systems of management. Traditionally, the management of forests has been dominated by the focus on timber. The literature on guidelines for their management exhibits this bias.*** While this may not affect

---

* See Prasad and Bhatnagar (1991a) and Gupta (1991) for details on the overextraction of this and other NTFPs.

** See Gupta (1991).

*** See, for instance, the review of this literature in Newman (1988), where rules for optimal rotation are discussed extensively. Campbell (1994) also discusses the significance of minor forest products in the management of forests.

the management of temperate forests significantly, multispecies tropical forests are affected in a more crucial manner. The significance of non-timber goods and services provided by these forests** has, however, been highlighted by recent studies. It is important to see the returns to be obtained from investment in the natural regeneration of a hectare of forest land on two alternative assumptions: that the benefit stream is conceptualized as consisting of both timber and non-timber products and that it is comprised of NTFPs only.

This investigation has been attempted in this section for degraded forest land containing sal rootstock and located in the laterite region of Raipur. The internal rates of return from forest management are obtained under alternative situations. The assumption is also made that some investment is necessary to enable natural regeneration. Direct costs include the cost of land management to contain the spread of weeds such as lantana and labor required for supplemental planting, application of fertilizer and insecticide, irrigation in initial years, and watch and ward throughout the project life. The management pattern envisaged contains some relatively new ingredients, such as complementary investment in socio-economic investment. This investment in socio-economic development of adjoining villages is aimed at creating indirect benefits for forest management by containing biotic pressure.***

Apart from the above investment, which amounts to about 30% of the total per-hectare investment, it is also somewhat traditional in its dependence on watch and ward and plantation activity. The benefit stream in the first situation in which both timber and non-timber benefits accrue has the following components:

- Fuel wood obtained after thinnings throughout the lifetime of the project
- Saw timber class 2 and large poles obtained in the 30th, 40th, and 50th years as per the specification of the model
- Bamboo after the 10th year, once every four years
- Small poles in the 10th, 20th, 30th, 40th, and 50th years
- NTFPs valued as per the values obtained in the present study to accrue after the 10th year on a sustainable basis every year

---

* See, for instance, the review of this literature in Newman (1988), where rules for optimal rotation are discussed extensively. Campbell (1994) also discusses the significance of minor forest products in the management of forests.

** See, among others, the work of Peters et al. (1989), who value the goods and services that a standing tropical forest provides, and Hartman (1976), who provides an analytical framework for valuing the services from a standing forest relative to those obtained from harvesting it.

*** The cost and benefit pattern is the same as that suggested for the development of an integrated forestry project for Madhya Pradesh (see Madhya Pradesh Integrated Forestry Project [1993]).

In the second case, where only non-timber benefits are considered, saw timber (small and large poles) is not included in the series of benefits accruing from the project; also, to study the sensitivity of the management model to time horizons, two alternative gestation periods of 40 and 50 years are considered. Harvesting costs are taken into account in the years when produce is harvested at the following rates:

- For timber, the cost is Rs240 (US$8) per cubic meter of saw timber harvested and transported to the forest gate.
- For bamboo, it is Rs600 (US$20) per hectare.
- For fuel wood, it is Rs600 (US$20) per hectare.

The above figures imply that the handling costs up to the forest gate are being included in total costs. As in all exercises of project evaluation, the prices taken are those prevailing at the beginning of the period; in this case, prices in Raipur are taken for all products. Again, to check for sensitivity, NTFPs are also evaluated at collectors' prices in one set of results.

It is important to keep in mind that the internal rates of return (IRRs) obtained are based on the direct benefits and costs only. The extent of underestimation seems to be higher in the case of benefits. This is because indirect benefits corresponding to costs incurred on socio-economic investments have not been accounted for.

The streams of benefits and costs analyzed in the project are given in Figures 6.1 and 6.2. Table 6.8 gives the IRRs obtained in the different alternative situations studied. When all benefits are valued at Raipur prices and when timber and non-timber accruals are both taken into account, the IRR is in the neighborhood of 20%. Further, it does not seem to be sensitive either to a decrease in the gestation period to 40 from 50 years or to the prices at which NTFPs are valued (i.e., Raipur prices vs. collectors' prices). Interestingly, even when only NTFP benefits are considered, the IRRs are greater than 12% in all cases. In other words, it would be worthwhile to consider investing in the natural regeneration of this sal tract even if NTFP benefits were the only ones that accrued, provided, of course, that the social rate of discount is 12% or less.* In the case of a tropical or subtropical forest, the third alternative of timber alone is not a viable alternative, keeping in mind ecological sustainability. One policy conclusion can be that although a trade-off exists, an ecologically sustainable combination of harvesting timber and non-timber products should be the aim.

---

* It is relevant to point out here that some analysts would maintain using lower rates of discount in natural resource planning. It was argued in earlier sections that giving a higher relative weight to preservation would have the same impact. Tampering with a national-scale parameter such as the rate of discount can have other consequences; see Markyanda and Pearce (1991).

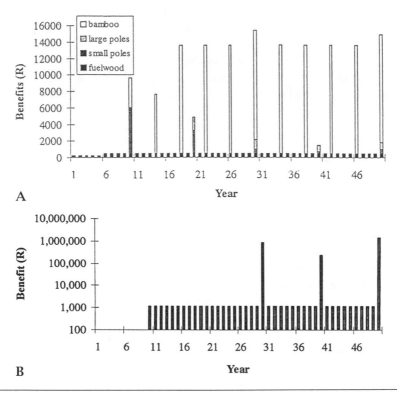

Figure 6.1 Estimated annual benefit stream, in rupees (Rs), from one hectare of regenerated sal forest for 50 years into the future (see text for explanation). (A) Annual benefits from non-saw timber. (B) Annual benefits from saw timber.

Non-Instrumental Value: Biodiversity Valuation

The different notions of value dealt with so far are all linked to instrumental value. However, NTFPs are also known to constitute a storehouse of biodiversity.* In that capacity, they contribute to the sustenance of the complex web of inter-actions that support human, plant, and animal life. Valuation of such a contri-bution is not easy. The natural environment is an extremely complex global system operating within a space–time continuum, and many linkages are not even understood in their entirety.** Although empirical estimates of existence

---

* This is all the more true of the large array of medicinal plants to be found in tropical deciduous forests.

** While this standpoint is often taken by ecologists, economists are now beginning to introduce such issues in their discussions; see, for instance, Terhal (1992). The interface of economics and ecology is also concerned primarily with this area; see Pearce and Moran (1994) and chapters in Wilson (1988).

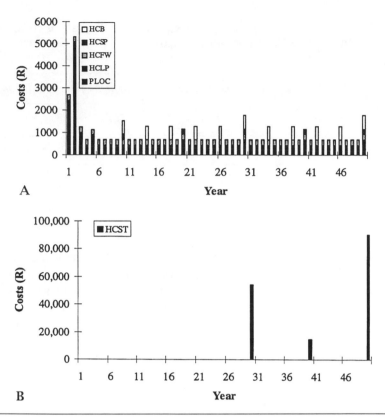

Figure 6.2 Estimated annual costs, in Rupees (R), of regenerating one hectare of sal forest and harvesting forest products for 50 years into the future (see text for explanation). Abbreviations: HCB = harvesting costs of bamboo, HCSP = harvesting costs of small poles, HCFW = harvesting costs of fuel wood, HCLP = harvesting costs of large poles, PLOC = plantation and operation costs, HCST = harvesting costs of saw timber. (A) Annual costs for non-saw timber. (B) Annual costs for saw timber.

value are supposed to capture this aspect,* it is difficult to say in an empirical sense what the component of bequest- or future-use value in these estimates is. The dividing line between option value, quasi-option value, and existence value is rather fine in an empirical situation.

It is possible to agree at an abstract level that one must look beyond a purely anthropocentric view of the scheme of the universe in order to conceptualize a

---

* See estimates contained in Brown (1992) for carbon sequestration of forests and in Godoy et al. (1993) for non-timber forest products.

Table 6.8 Natural Regeneration of Degraded Sal Forests in Raipur: Expected Direct Benefits (internal rates of return in percent) from 1 Hectare

	Benefits from timber and NTFPs		*Benefits from NTFPs only*	
Gestation (years)	*NTFPs valued at Raipur prices*	*NTFPs valued at collectors' prices*	*NTFPs valued at Raipur prices*	*NTFPs valued at collector's prices*
50	20.4	20.3	13.12	13.10
40	20.35	20.37	12.8	12.50

purely non-instrumental value. However, most attempts at valuation of biodiversity find it difficult to proceed beyond an anthropocentric view of the environment if the focus is on issues of policy. Indeed, it may even be counterproductive to do so. It is within such a context that we shall view the issue of biodiversity valuation. Even within such a framework, there is lack of agreement on the unit of measurement of biodiversity. The unit may be the number of species, the keystone species, the genetic information preserved *ex situ,* or the ecosystem. The technique and cost of biodiversity valuation would differ according to the unit considered relevant.

If it is considered that the number of species in existence that is of essence, a social welfare function that gives a value to this number becomes relevant. This view of biodiversity maintains that it is embodied in the existence of opportunities available for development envisaged for the future.* Further, if sustainable development requires that this set of opportunities with respect to the future be an increasing one, and if an increase in the components of a subset of natural capital is a necessary condition for this to happen, biodiversity as measured by the number of species in existence at any point in time (both animal and plant) becomes an independent element of the implicit social welfare function.

The argument for the above approach lies in the axiom of choice wherein the existence of a larger number of elements in a choice set gives society a higher level of welfare, irrespective of whether that choice is exercised or not. Such preference functions have been a part of the accepted preference pattern in societies and have been analyzed in the literature.** The implication of such an

---

* In recent contributions, Perrings (1992) and McNeely (1992) have adopted such an approach.

** See, for instance, Sen (1991), who based one view of development on the notion of the expansion of the choice frontier.

approach for natural resource policy is that any option that reduces the number of species is inferior to one that does not. Preservation of biodiversity (viewed as the number of species) becomes a necessary condition for development to be sustainable. It may be contested that this is a very stringent condition. It is also one that is rooted in an anthropocentric view of the development process and emerges from social preference functions that are a part of welfare economics. This requirement for sustainable development has far-reaching implications for the management of forests, in particular in the context of NTFP production. The requirements may seem unrealistic in situations of low present income where a trade-off between present use value, future use value, and biodiversity value may be necessary. It is here that the notion of biodiversity as part of a global preference function may have a crucial role to play. If sustainable development is indeed an overriding international concern, a global approach to methods of biodiversity preservation will have to be adopted. Geographical locations that harbor large amounts of biodiversity are fairly well identified. It is known, for instance, that biodiversity increases in general with a decrease in latitude. If it is also agreed that there is a pronounced global dimension to the value of biodiversity, then it is significant for the human race per se. There is, therefore, an implicit non-instrumental aspect to its valuation. Capturing this global value is of the essence in this area. Even while details of measurement are being debated, the creation of global mechanisms for the preservation of value should proceed apace, however incomplete or approximate the measurement aspect may be.*

Conclusions

Some NTFPs from among the range available in the forests of Raipur are selected for focusing on issues relating to valuation and pricing. The study also evaluates the role of market and non-market institutions in ensuring a sustainable supply of such products. Starting from a user valuation-oriented approach, the focus is on the manner in which forest communities as users view NTFPs. Primary data from surveys conducted in the district are used to understand the role of the market and of different market channels in the local economy. It is found that the collection and sale of NTFPs are important sources of secondary income for rural households in the region. On an average, 40% of household

---

* See, for example, the discussion in Pearce and Moran (1994) on the ways in which exotic capital can be preserved, either through an increase in consumer demand of a certain variety, or through large firms seeing their prospects of future profit, or yet again through international organizations, such as the Global Environment Facility.

income is contributed by them. This collection has a seasonal aspect, though. Some products such as tendu leaves add considerably to collectors' incomes at certain times of the year. A study of the markets in which these products are sold reveals inequities in the market structure. A collector's price as a percentage of the sale price varies from 26 to 85%. A number of alternative market channels exist. However, they differ considerably in the percentage of the retailer's price that the collector is able to appropriate. Available data seem to indicate, however, that this percentage is higher for high-value products.

Interventions in markets for NTFPs have existed as (1) nationalization and (2) assuring supply to the less privileged. Both these kinds of interventions are of some value. The first may protect the collector from year-to-year variations but may not necessarily increase the proportion of labor income to final price. Assuring a preference in supply to certain sections has been another form of intervention for assuring equitable access. It seems to be superseded by agreements to supply to other sources of demand offering higher prices. On balance, moving toward a situation of more integration with markets seems to be the preferable policy option. Making it possible for market channels to process products more smoothly turns out to be the best kind of intervention. It can be added here that while market channels throw light on the efficiency with which collected produce is brought to the market, there is little they can reveal with respect to the sustainability of the amount collected. As noted above, however, markets do have limitations and social interventions are still necessary under certain circumstances.

When sustainability cannot be ensured, the NTFP acquires the nature of an exhaustible resource as a consequence of overextraction, and option value becomes relevant. Values attached to abstaining from use in the present are estimated using two alternative methods. This analysis yields a range of somewhat hypothetical opportunity costs of present use or preservation values. The implications of policy decisions taken by the Indian government in the past with respect to the exploitation of certain NTFPs are examined in the context of these hypothetically generated values. It is found that they sometimes reveal a preference for the future higher than expected by the norm of the social rates of discount.

Techniques of forest management focusing on a range of NTFPs are known to promote sustainability. This study shows that they yield rates of return of 12 to 13%, thereby supporting the hypothesis that ecologically sound management is economically efficient as well. In conclusion, it can be stated that enabling markets in NTFPs to function more smoothly is imperative from the point of view of collectors. Further, managing forests for NTFPs is economically viable, although some form of intervention may be necessary to ensure sustainability.

This may be in the form of government- or community-determined constraints on rates of extraction.

Appendix

This appendix describes the NTFPs of greater significance in Raipur, including how and when they are used.

Bamboo

The two most important bamboo species of India are *Dendrocalamus strictus* and *Bambusa arundinacea*. The latter is common in South Raipur district. Bamboo production in Madhya Pradesh in 1989–90 was 333,000 national tonnes. It is a most versatile product and has multiple uses: commercial, industrial, and as rural construction material. It is used as a major raw material in the newsprint and rayon industries. In addition to commercial and industrial uses, numerous articles of daily use, such as fences, handicraft items, tool handles, tent poles, toys, musical instruments, etc., are made of bamboo. It is the main raw material for basket and wicker work. For the poor, bamboo is indispensable as raw material for handicrafts, house construction, agricultural implements, and basket weaving. A decline in bamboo production over time has been noted in this region. This decline has primarily been on account of gregarious flowering and consequent drying of bamboo clumps. Though a large number of germinants come up, they are not allowed to establish due to intense biotic pressure. Excessive grazing, recurring fires, and overexploitation of bamboo forests make establishment difficult. Study of the flowering pattern of bamboo therefore assumes great importance in managing the forests optimally and for providing adequate and appropriate treatments for retrieval of bamboo forests.

Leaves

Leaves of a number of trees, shrubs, herbs, and climbers have great economic value. Leaves of *Diospyros melanoxylon* are used as beedi wrappers and are thus an important item of commercial value. Leaves of *Butea monosperma* and *Bauhinia vahlii* are used for making leaf plates. Leaves of some climbers are used as spices. Use of leaves as roof thatching material also enhances the value of many forestry species. Apart from being commercially valuable, leaves of many forestry plants also provide year-round employment to local inhabitants, in particular tribal people and other forest dwellers.

Diospyros melanoxylon is found in sal forests, often replacing sal in areas having poor regeneration potential. The manufacturing of beedi from these leaves (known locally as tendu) has become a well-established and important cottage industry in Madhya Pradesh. It is estimated that about 60% of the total production of the country comes from this state, and 75% of this production is consumed within the state for manufacture of bidies. In 1990, the collection of tendu leaves in Raipur was about 300,000 standard bags and comprised about 5% of the total collection in Madhya Pradesh.

Bauhinia vahlii or mahul is a gigantic creeper, one of the most abundant of the Indian climbing *Bauhinia* species. Most of the collection is reported from the sal region. Although it is a fiber-producing climber, it is usually looked upon as a pest, due to the damage it can cause to healthy trees. The leaves, however, are used extensively as leaf plates and cups and also in "pan" shops as wrappers. The average yearly collection of mahul in Raipur in 1990 was 6093 quintals.

Tree-Based Oilseeds

In addition to the annually cultivated oilseed crops, 86 different oilseed-bearing trees are found in India. These occur naturally in the forests and are principally managed for the timber they yield. The oils obtained from forest trees are of varying commercial importance. Some of these oils or "butters" have already established a good market for themselves, but the majority are of local importance only. The most common oilseeds are sal seed, mahua, karanj, and kusum palas bhilwa.

The cotyledons of sal seed yield the well-known sal butter. The seeds are husked and boiled and the oil is skimmed off. The oil soon solidifies to a white butter and is used for cooking and lighting and for adulterating of ghee. It is a suitable confectionery fat and may also be used in soap making. The seeds are sometimes eaten whole, especially in times of famine.

Mahua or *Madhuca latifolia* is a large evergreen tree with numerous branches. Mahua oil is hard in nature and constitutes 35% of the seed. It is used in the production of soaps. However, this fat, properly refined, can be used for cooking in confectionery and chocolate making. Refined oil finds use in the manufacture of lubricating greases.

The karanj tree is primarily used for its seeds. The oil content of karanj is about 27%. The oil is chiefly used for leather tanning, soap making, lubrication, and in medicine. Both the seed and oil possess remarkable medicinal properties. The oil cake is a good fertilizer.

Kusum is a large deciduous tree and bears minute yellow-green flowers. The fruit or the hard-skinned berry contain one or two ellipsoidal seeds with a brownish coat. These trees serve as hosts for lac insects and are generally found in compact blocks; this facilitates the collection of seed. It is utilized by the soap industry in the unorganized sector. It has been used for a long time in hairdressing and in some medicinal preparations for skin diseases, rheumatism, and headaches. The tree is lopped for fodder, and flowers yield a dye. Raipur and Kanker districts in Madhya Pradesh are the most significant areas, contributing 96% of the total collection of the state. In Raipur, the average yearly collection in 1982–85 was about 1270 quintals.

Palas or *Butea monosperma* is commonly known as "the flame of the forest." It is a medium-sized tree found throughout India. Palas is used extensively for lac propagation. The oil finds good use in soap making. Raipur showed a yearly collection of 213 quintals of palas during 1982–85.

The oil of neem is used in soap manufacture. Practically every part of the tree is bitter and is used in indigenous medicine. The bark is a good bitter tonic and astringent and is also useful in skin diseases. Among the oil cakes, neem cake is sometimes preferred for manuring certain crops, such as sugarcane.

Tans and Dyes

Among the bark tans, the bark of Babul is the cheapest source of tannin suitable for heavy leathers. The tree yields a charcoal fuel, fodder, and small timbers for agricultural implements. Of the leaf tans, dhamada or *Anogeissus* is universally distributed in all forest types. The fruit of the tree *Terminalia chebula*, commercially known as chebulic myrobalan, is an important tanning material. Madhya Pradesh accounts for 75% of the total production in the country; Raipur is an important market. Production has been fluctuating, partly due to the felling of these trees as a consequence of clearance for agriculture. In 1989–90, the production was 114,000 quintals.

Lac

Generally known as shellac, lac is used for a variety of purposes in plastics, electrical goods, adhesives, leather, wood finish, printing, polish and varnish, ink, and a number of other industries. It is also a principal ingredient of sealing wax. The two important strains of lac insect are *Rangeeni* and *Kusumi*. Its annual production in Raipur during 1981–82 was 1780 quintals.

Edible Products

Natural forests supplement the food supplies for human beings, particularly tribal people. Numerous forest fruits and seeds, flowers, rhizomes, tubers or roots, barks, honey, and wax are consumed by people not only during periods of food scarcity and famine but also in normal times. Some examples are given here.

Chironji (*Buchanania lanzan*) is an important tree species and is found in mixed dry deciduous forests of Madhya Pradesh. The tree has economic importance for the edible fruit it yields. The fruit is eaten by local people, and kernels are extracted and dried for sale in the market. Raipur shows an average collection of 60 quintals of chironji.

The aonla tree (*Emblica officinalis*) is common to the mixed deciduous forests. The fruit is green when tender, changing color to light yellow or brick red when it matures. However, the fruit is better known for pickles, preserves, and jellies. Due to various preparations made from fresh aonla, it is a valued fruit with a ready market. It is also collected for use in the preparation of ayurvedic tonics and medicines. The average yearly collection is assessed at 1463 quintals in Raipur.

Honey forms a natural nutritious food for the rural people. It is used for medicinal purposes. The average yearly production of honey in Raipur is 82 quintals.

There are two species of mahua: *Madhuca indica* or *latifolia* and *Madhuca longifolia*. Both are valued for their flowers and fruits. The fleshy, cream-colored corollas of its flowers are eaten raw or cooked by poor people. They are a rich source of sugar, vitamins, and calcium. These flowers are known to provide a source of sustenance to the poor in times of stress, such as drought. In normal times, a market exists for it as it is used in the brewing of a local liquor.

Gums and Resins

Some plants yield only gum, others yield only resins, and still others yield both gum and resins. Gums find uses in a variety of industries, such as food, textile, paper, dyeing, and adhesives. They are also used in the cosmetics, paint, ceramic, and ink-making industries. Commercial gums enter the market in the form of dry exudation.

Medicinal Plants

Dependence of humankind on medicinal plants is second only to dependence on food for life. The Ancient Hindu texts Rigveda and Sushruta Sahinta mention their importance in India, and China and Greece are also pioneers in this field. About 2000 plant species are medicinal, of which about 1200 species (25% of the Indian flora) support about 5000 manufacturers of indigenous drugs in the Indian system of medicine. Nearly 200 of these units make about 2000 preparations in different parts of the country. It is believed that about 80% of medicinal plants come from forests.

References

Arrow, K.J. and Fisher, A.C. (1974) Environmental preservation, uncertainty and irreversibility. *Quarterly Journal of Economics,* 88: 313–319.

Bawa, K. and Godoy, R. (1993) Introduction to case studies from South Asia. *Economic Botany,* 47: 248–250.

Bishop, C.R. (1982) Option value: an exposition and extension. *Land Economics,* 58: 123–127.

Biswas, P.K. (undated) Non-Nationalized Minor Forest Products and Tribal Economy: A Case Study in Madhya Pradesh. Working Paper of the Indian Institute of Forest Management, Bhopal, India.

Brown, K. (1992) Carbon Sequestration and Storage in Tropical Forest. Centre for Social and Economic Research on the Global Environment, Discussion Paper 92. University of East Anglia and University College London, Norwich and London, U.K.

Campbell, J.Y. (1994) Changing Objectives, New Products and Management Challenges: Making the Shift from "Major vs. Minor" to "Many" Forest Products, paper presented at a seminar on forest products, Coimbatore, India.

Centre for Minor Forest Products, Newsletter (different years), Dehradun, India.

Centre for Monitoring the Indian Economy (1993) Profiles of Districts, CMIE, M-79, M Block Market, Greater Kailash-II, New Delhi-48.

Chakravarty, R. and Prasad, R. (1989) Forestry-based tribal development: an approach. *Journal of Tropical Forestry,* 5: 234–241.

Chopra, K. (1993) The value of non-timber forest produce: an estimation for tropical deciduous forests in India. *Economic Botany,* 47: 252–257.

Das, J.K. (1992) Valuation of timber and NTFPs, paper presented at the Workshop on Methods for Social Sciences Research on Non-Timber Forest Products. University of Kesetsart, Bangkok, Thailand, May 18–20.

Edwards, D.M. (1993) The Marketing of Non Timber Forest Produce from the Himalayas: The Trade between East Nepal and India. Rural Development Forestry Network, Overseas Development Institute, London, U.K.

Godoy, R. (1993) The Effects of Income on the Extraction of Non-Timber Forest Products among the Sumu Indians of Nicaragua: Preliminary Findings. Manuscript. Harvard Institute for International Development, Cambridge, MA.

Godoy, R. and Bawa, K. (1993) The economic value and sustainable harvest of plants and animals from the tropical forest: assumption, hypothesis and methods. *Economic Botany,* 47: 215–219.

Godoy, R., Lubowski, R., and Markhandya, A. (1993) A method for the economic valuation of non-timber forest products. *Economic Botany,* 47: 220–233.

Government of Madhya Pradesh (1993) Madhya Pradesh: State of Forests. Forest Department. State Forest Research Institute, Jabalpur, India.

Gupta, B.N. (1991) Status of non-wood forest products in India, paper presented at a Regional Expert Consultation on Non-Wood Forest Products at the FAO Regional Office, Bangkok, Thailand, Nov. 5–8.

Hartman, R. (1976) The harvesting decision when the standing forest has value. *Economic Enquiry,* 14: 213–216.

Madhya Pradesh Integrated Forestry Project (1993) Consultants Reports. Bhopal, India.

Madhya Pradesh State Minor Forest Products Trading and Development Cooperation Federation Limited (1991) Annual Reports, Vol. 3 (IV). Vikas Bhawan, Bhopal, India, March 30.

Markyanda, A. and Pearce, D.W. (1991) Development, the environment and the social rate of discount. *The World Bank Research Observer,* 6: 137–152.

Marothia, D.K. (1992) Cooperative management of minor forest products in Madhya Pradesh: a case study, paper presented in the Workshop on Cooperatives in Natural Resource Management at the Symposium on Management of Rural Co-operatives, Dec. 7–11, 1992. Institute of Rural Management, Anand-388 001, India.

Marothia, D.K. and Gauraha, A.K. (1992) Marketing of denationalized minor forest produce in a tribal economy. *Indian Journal of Agricultural Marketing,* 6: 157–160.

McNeely, J.A. (1992) Biodiversity in India: some issues in the economics of conservation and management. In: *The Price of Forests.* A. Agarwal, Ed. Centre for Science and Environment, New Delhi, India, pp. 125–131.

National Centre for Human Settlement and Environment (1987) Documentation on Rests and Rights, Vol. 11, Document B-7/16A. Safdarjung Enclave, New Delhi, India.

Newman, D.H. (1988) The Optimal Forest Rotation: A Discussion and Annotated Bibliography. A Study Sponsored by U.S. Department of Agriculture, Forest Service, Southeastern Forest Experiment Station, Asheville, NC.

Operation Research Group (1993) Comprehensive Evaluation of Forest Based Cottage Industries in Madhya Pradesh, Vol. II. Submitted to Chief Conservator of Forests (Task Force), Madhya Pradesh Forestry Project, Bhopal, India.

Pearce, D.W. and Moran, D. (1994) *The Economic Value of Biodiversity.* Earthscan Publications, London, U.K.

Pearce, D.W. and Turner, R.K. (1990) *Economics of Natural Resources and the Environment.* Harvestersheaf, U.K.

Perrings, C. (1992) Biotic diversity, sustainable development and natural capital, revised version of a paper presented at II Biennial Meeting of the International Society for Ecological Economics, Stockholm, Sweden.

Peters, C.M., Gentry, A.H., and Mendelson R.O. (1989) Valuation of an Amazonian rain forest. *Nature,* 339: 655–656.

Prasad, R. and Bhatnagar, P. (1991a) Socioeconomic Potential of Minor Forest Produce in Madhya Pradesh. Report of the State Forest Research Institute, Jabalpur, India.

Prasad, R. and Bhatnagar, P. (1991b) Wild edible products in the forests of Madhya Pradesh. *Journal of Tropical Forestry,* 7: 234–245.

Prasad, R. and Pandey, R.K. (1987) Survey of medicinal wealth of central India. *Journal of Tropical Forestry,* 3: 287–297.

Prasad, R. and Pandey, R.K. (1992) An observation on plant diversity of sal and teak forests in relation to intensity of biotic impact at various distances from habitation in Madhya Pradesh. *Journal of Tropical Forestry,* 8: 264–270.

Reddy, V.R. (1994) User Valuation of Renewable Natural Resources: Some Methodological Issues. Mimeographed report of the Institute of Development Studies, Jaipur, India.

Sen, A.K. (1991) Welfare, preference and freedom. *Journal of Econometrics,* 50: 15–19.

State of Forest Report (1991) Government of India, Forest Survey of India. Ministry of Environment and Forest, Dehradun, India.

Terhal, P. (1992) Sustainable development and cultural change. In: *Environmental Economy and Sustainable Development.* J.B. Opschoor, Ed. Wolters-Nootdhoff, Amsterdam, Netherlands, pp. 129–142.

Wilson, E.O., Ed. (1988) *Biodiversity.* National Academic Press, Washington, D.C.

Poverty Alleviation, Empowerment, and Sustainable Resource Use: Experiments in Inland Fisheries Management in Bangladesh

7

Ana Doris Capistrano
Ford Foundation, Dhaka, Bangladesh*

Mokammel Hossain
Bangladesh Department of Fisheries, Dhaka, Bangladesh

Mahfuzuddin Ahmed
International Center for Living Aquatic Resources Management, Manila, Philippines

Abstract

Complex linkages between environmental resource degradation and powerlessness, especially as they relate to common property resources in developing countries, argue for operational approaches to sustainable development grounded in poverty alleviation, empowerment, and intergenerational equity. Since government policies provide the context within which resources are accessed, allocated, and used, they play a crucial role in promoting or obstructing the process.

* Present address: Ford Foundation, New Delhi, India.

1-57444-077-2/97/$0.00+$.50
© 1997 by CRC Press LLC

This chapter reviews Bangladesh's policy experiments in managing its vast inland fisheries. It focuses on an action-research project with the government in support of a policy attempting to simultaneously promote social equity, poverty alleviation, and sustainable use of fishery resources. It discusses lessons learned, and new approaches taken as a result, and notes continuing challenges.

Introduction

Until recently, environmental conservation and economic development have been viewed as competitive policy objectives. Evidence now suggests that they can be mutually consistent (World Bank, 1987; Pearce, 1988; Daly and Cobb, 1989). As the single-minded pursuit of economic growth gradually gave way to a movement toward "sustainable development" over the past two decades, the economic dimension of development has come to be seen as only part, albeit a very important part, of an environmentally sound socio-cultural, political, and institutional transformation (Barbier, 1987; ANGOC, 1989; Dietz and Straaten, 1992; Pezzey, 1992).

However defined, sustainable development rests on an underlying ethical foundation of intergenerational equity. This is clearly reflected, for example, in the well-known World Commission on Environment and Development (WCED) definition as development that "meets the needs of the present generation without compromising the ability of future generations to meet their own needs" (WCED, 1987, p. 43). Attempts to more "sustainably" manage environmental resources have indeed highlighted the centrality of equity, not just between generations but also within generations. Better understanding of the complex linkages between environmental degradation on the one hand and poverty and powerlessness on the other (WCED, 1987; Jazairy et al., 1992; Oodit and Simonis, 1992; Boyce, 1994), especially as they relate to common property resources in developing countries, now argue for the adoption of operational approaches to sustainable development that are grounded equally in poverty alleviation, empowerment, and intragenerational social equity. Since government policies provide the context within which resources are accessed, allocated, and used, they play a crucial role in fostering or inhibiting sustainable development.

This chapter reviews Bangladesh's policy experiments in managing its vast inland fisheries. It focuses on an action-research project with the government in support of a policy attempting to simultaneously promote social equity, poverty alleviation, and sustainable resource use. It discusses lessons learned, and new approaches taken as a result, and notes continuing challenges.

Rights, Power, and Common Property Resource Management

Common property resources such as fisheries are a significant source of subsistence and livelihood, especially for poor people in developing countries. The term "common property resources" is used here to refer to resources used by individuals under a variety of property rights arrangements. The poor, although often regarded as the proximate agents of resource destruction, are also usually its first victims and thus have a major stake, perhaps more so than other users, in the management and conservation of resources.

Sustainable management of common property resources, especially fisheries, is complicated by their defining characteristics: (1) the difficulty of excluding other users and (2) the use by one agent subtracts from the amount of the resource available to others (Berkes, 1989; Bromley, 1991; Charles, 1994). Typically, common property resources are owned, though not necessarily effectively managed, by the state. Privatization, the classic Hardinian prescription to combat overexploitation (Hardin, 1968), is not always possible or workable given the generally high transaction costs associated with enforcing private property rights. In many cases, common property resources are best left in the realm of the commons and regulated by collective action involving their various stakeholders.

To be effective in protecting the integrity of common property resources, regimes for their management require well-defined, clearly understood, and enforceable property rights and responsibilities (Berkes, 1989; Ostrom, 1990; Ostrom and Gardner, 1993). A growing body of literature shows that sustainable resource use develops when a particular group of stakeholders has both control and responsibility for the resource (Gadgil and Berkes, 1991; Dyer and McGoodwin, 1994). Furthermore, collective arrangements for resource management tend to work better in small groups with similar needs and shared norms and patterns of reciprocity within clearly defined boundaries (Goulet, 1989; Ostrom, 1990). Such arrangements tend to develop among foresighted individuals who anticipate future gains from cooperation (Seabright, 1993; Bardham, 1993).

Institutions for cooperative resource management work more successfully when they are embedded in a context in which collective action has worked in the past. Moreover, well-functioning management regimes are likely to be (1) equitable (i.e., enjoy a shared perception of fairness), (2) efficient (i.e., have a minimum or absence of disputes and require limited effort to maintain compliance), (3) stable (i.e., have a capacity to cope with progressive changes through adaptation), and (4) resilient (i.e., have a capacity to accommodate surprise or sudden shocks) (Gibbs and Bromley, 1989).

Obviously, the costs and benefits to individuals of participating in these management institutions vary and determine the extent, mode, and manner of their involvement. Participation can strengthen the claims to resources of poor or marginalized individuals who, on their own, may have inequitable relationships with local institutions and little access to channels of decision making. However, when benefits from management are defused, have long gestation periods, or are made uncertain by technical or social factors, unless there are immediate benefits to be gained, the poor are less able and less likely to participate.

Although privatizing common property resources may be problematic, it is often possible to privatize and assign rights over these resources to individuals or collectives to use, manage, and exclude others. Assignment of rights may be explicit or implicit and may be *de jure* (by law) or *de facto* (in actual fact). Use rights to a resource would include right of access (i.e., right of entry to a defined physical property) and right of withdrawal (i.e., right to obtain the products of a resource, for example, to catch fish). Management right refers to the right to regulate internal use patterns and transform the resource by making improvements.* Exclusion right pertains to the right to determine who will have access and how that right may be transferred (Schlager and Ostrom, 1992).

To a large extent, the distribution of these rights correlates with the relative distribution of power and influence among the resources' various stakeholders. The greater the inequity in the distribution of rights and power, then, all else being equal, the greater the level of environmental resource degradation is likely to be (Boyce, 1994). Furthermore, contrary to Coase's (1960) theorem, property rights allocation that confers differential bargaining power to negotiating parties does affect the levels of resource use and the consequent environmental impacts (Pezzey, 1992).

Conceivably, a reallocation of property rights in favor of disempowered stakeholders (e.g., poor women, marginalized minority populations) may not

---

* If the ecological health of a resource were the only consideration, management intervention would have dubious results in the case of resources in a pristine state. However, if other objectives such as enhancing economic productivity come into play (e.g., increasing proportion of commercially valuable biomass out of total biomass produced, such as in agriculture or forestry), then judicious management presumably would balance competing demands on the resource. These definitions of rights come directly from Schlager and Ostrom (1992), and although they are perhaps deficient in some ways, conceptually they represent a quantum leap over the simplistic property rights definitions in popular use until a few years back.

only help alleviate poverty and enhance social equity but also engender more efficient resource use and conservation. Bangladesh's private voluntary development organizations (commonly referred to as non-governmental organizations or NGOs) have demonstrated that by incorporating grass-roots mobilization, environmental consciousness raising, appropriate technology transfers, and policy advocacy into their programs for economic and social empowerment, it may be possible to make the goals of equity, productivity, and resource conservation more consistent. Where applicable, this approach promises, and has in fact yielded, significant payoffs (Lovell, 1992; Capistrano, 1996). However, as the following experience with Bangladesh's inland open-water resource management would suggest, it also faces myriad challenges.

Bangladesh's Fisheries: Overview

Bangladesh, a land created by rains and rivers, is a floodplain of two of the world's greatest bodies of water, the Ganges and the Brahmaputra (Figure 7.1). Its 4 million ha of inland open waters (freshwater marshes, oxbow lakes, natural depressions, rivers, and estuaries) and almost 147,000 ha of closed water ponds are among the world's richest and most complex fisheries.* They provide habitat for 260 species of fish, more than 20 species of shrimps, and numerous other species of plants and animals (Ali, 1991; Nishat et al., 1993). The monsoon rains that typically flood over half of the country's land area of about 144,000 km^2 regulate fish migration, reproduction, and growth, which are essential to the productivity of the fisheries and to the sustenance of communities they support (ISPAN, 1993).

Fish–People Interactions

In this poverty-stricken and densely populated country of 120 million people, fish is second only to rice as a source of food. It represents 80% of animal protein consumption (ISPAN, 1993). About two million people are directly employed in fishing and another ten million in fish marketing and processing. From 75 to 85% of all rural households engage in seasonal subsistence fishing. Using a variety of fishing gear or just their bare hands, men, women, and children tap a diversity of niches in the floodplain and open waters. Contributing 4%

---

* Open water fisheries are those that become part of a single integrated fishery during the monsoon season. They may be either flowing or standing waters. Closed water fisheries, on the other hand, are those with minimal or no connections with rivers or floodplain and hence remain distinct even during the monsoon season.

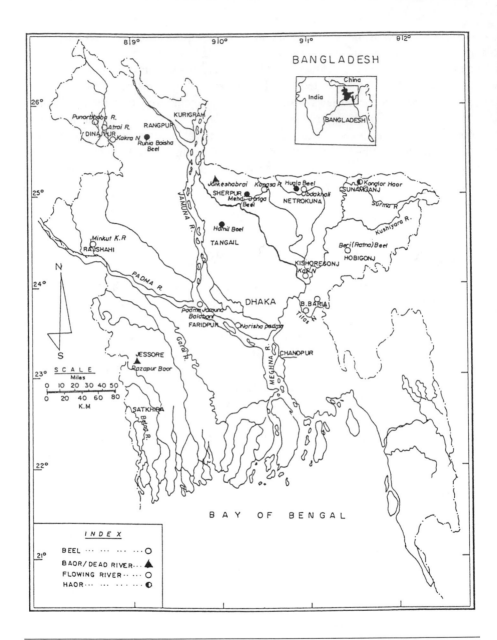

Figure 7.1 Map of Bangladesh showing freshwater fisheries under IMOF Project (Phase II).

of national output and 11% of export earnings, fishery is indeed a key sector in Bangladesh's economy and an important safety net for its poor (Minkin and Boyce, 1994). Perhaps nowhere else is the well-being of so many so closely intertwined with the waters, the fisheries, and the delicate balance that maintains them.

For centuries, Bangladesh's inland fisheries have adapted to the natural stresses of recurrent flooding, siltation, land formation, and low levels of human predation. Now, however, they are being overwhelmed by the accumulated stresses from rapid expansion of human activity: poorly designed roads and haphazardly planned flood control and irrigation structures which destroy fish habitat and interfere with fish movement and reproduction, wetlands conversion for agriculture and commercial land use, pollution from agricultural chemicals and industrial waste, and overfishing by a growing population using more efficient gear (Khan, 1993; Rahman, 1993; Ahmed, 1995b). These fisheries are also subject to the pervasive influence of a policy framework that has provided little incentive for fisheries conservation and now threatens to further undermine the incipient management arrangements that have developed in the last few years (Naqi, 1989; Siddiqi, 1989; Aguero and Ahmed, 1990).

Production and Productivity

In the absence of reliable data, statistics on Bangladesh's fisheries are at best informed guesses. Nevertheless, there is little disagreement that, although total production may have been fairly steady or even increasing in absolute terms (Figure 7.2), the catch per fisher has declined and the quality of the country's inland capture fisheries has deteriorated considerably in recent decades (TSS-1, 1992; Rahman, 1993; Ali and Fisher, 1995). Small fish species, classified as non-important "miscellaneous" species for the purposes of official statistics, now predominate with a 40% share of total catch (FRSS, 1994). Little is known about the biology and productivity of these "minor products" of the fisheries even though they are a vital part of the poor's diet and have been shown to provide more calcium, vitamin A, and essential nutrients compared to bigger, more expensive fish (ISPAN, 1993).

Commercially important carps and hilsha (*Hilsha illisa*), previously abundant species, now make up only 26 and 10% of catch, respectively. Between 1986 and 1992 alone, naturally spawned carp fingerling production is estimated to have declined by an average of 26% per year (FRSS, 1994). There are indications that certain species are endangered or have become locally extinct (Ahmed, 1995a).

Alarmed by obvious signs of degradation and attracted by the promise of advances in aquaculture technology, the Bangladesh government has turned its

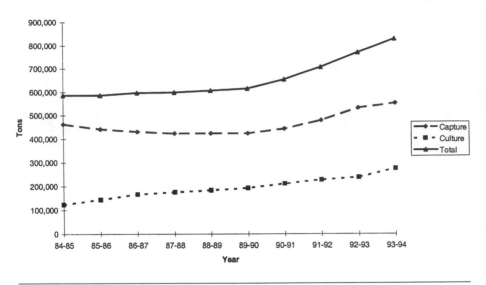

Figure 7.2   Inland fish production of Bangladesh, in metric tons, 1984–94. Source: FRSS, Bangladesh Department of Fisheries (1994).

attention to the fishery sector after decades of relative neglect. With massive funding from the country's international donors, in 1992 the government adopted a two-pronged strategy to counter productivity decline and increase export earnings from the sector: (1) restocking the floodplain with fast-growing, mostly exotic, carp species and (2) promotion of aquaculture (Ali and Fisher, 1995). Paradoxically, as attention turns to the fishery sector as a possible engine of economic growth, access of rural communities to fisheries resources is challenged more and more by both well-entrenched and newly emerging interest groups seeking to gain greater control over potentially valuable fishing grounds (Kremer, 1994a; Capistrano et al., 1994). Especially for the 40% of the country's population living in extreme poverty, continued access to these resources is under increasingly serious threat.

Fisheries Development Projects

Funding for fisheries development projects from 1991 to 1995 was estimated at US$332 million, $240 million of which was from external sources (TSS-1, 1992). A major portion of foreign funds came from the World Bank and the Asian Development Bank to support programs in floodplain restocking, programs which the government credits for helping increase production and fishing income (Ali and Fisher, 1995) but which have raised serious concerns about their ecological and equity implications (Kremer, 1994a,b; Minkin and Boyce, 1994; Naqi et al., 1994). A number of bilateral donors, notably the British Overseas Development Agency and the Royal Danish Embassy, support aquaculture technology development and extension for use mostly in closed water ponds but also in smaller open water bodies and rice fields (Wood and Gregory, 1990; Ali et al., 1993).

Bangladesh's NGOs, noted for their dynamism and innovation as well as size and scale of operations, have been quick to take advantage of opportunities in the fishery sector created by technology development and availability of donor funding. Many of the country's over 600 development NGOs implement fisheries projects as part of their poverty alleviation programs. NGOs' fisheries projects typically involve organizing homogeneous groups of poor people for aquaculture management on private and state-owned ponds, canals, lakes, and other bodies of water (FAP-17, 1993). Applying group mobilization, training, and social empowerment strategies successfully used in the selling of irrigation water by the landless poor (Wood, 1984; Wood and Palmer-Jones, 1991), NGOs work to secure use rights or contracts to manage fisheries for their groups (FAP-17, 1993; Wood and Gregory, 1990; Watanabe, 1993). The NGOs bear the transaction costs of negotiating with resource owners. Having secured a lease or management contract, the NGOs then enter into a range of co-management and profit-sharing arrangements with their group members in which they typically provide credit, marketing, and technical support services (Ahmed et al., 1995; Capistrano, 1996).

Evolution of Fisheries Management Policies

Fisheries in pre-colonial Bangladesh were traditionally managed as common property resources through complex systems of tenure evolved in and enforced by local communities. During the colonial period, however, laws passed by the British to maximize state revenue generation gave zamindars (feudal lords) proprietary rights of use, management, and exclusion over water bodies within their estates (Farooque, 1989). The zamindars collected a nominal tax in exchange for use rights to the fisheries which served, in effect, to regulate entry and harvest within sustainable limits.

When the East Bengal State Acquisition and Tenancy Act of 1950 abolished the zamindari system, the majority of the country's open water bodies, as well as 100,000 of its 1.7 million ponds, reverted to the state (Farooque, 1989). The Ministry of Lands (MOL), one of the most powerful government agencies in the country, currently has authority and proprietary rights over these state-owned water bodies, although 25 departments in 13 other ministries are also responsible for aspects of their management and regulation (Ahmed and Hossain, 1995). This includes the Department of Fisheries (DOF), a technical department under the Ministry of Fisheries and Livestock (MOFL), which is responsible for fisheries administration, management and conservation, extension and training, and regulatory enforcement. With the multiplicity of agencies involved, fisheries management is made more difficult by a lack of coordination and, unfortunately, is often held up by interministerial squabbles (Siddiqui, 1989).

In a virtual extension of colonial policy, the MOL has, since 1960, managed state-owned fisheries primarily to raise revenues and has leased segments of water bodies to the highest bidder for short-term (usually one to three years) periods. Deputy commissioners stationed in each of the country's 64 districts are responsible for overseeing on behalf of the MOL the leasing of water bodies and collection of revenues. As the government's chief executive in the district, the deputy commissioner has the power and control over fisheries resource allocation.

Although, according to government policy, cooperative groups of traditional fishers are supposed to get priority in lease allocation, in practice, leases usually go to an influential elite composed of middlemen, politicians, and moneylenders (Naqi, 1989; Siddiqi, 1989). Under this system, traditional fishers, who are mostly poor, low-caste Hindus (a minority population in predominantly Muslim Bangladesh), have lost significant use rights to valuable customary fishing grounds, especially during the winter season, when returns on fishing are highest. Bengali Muslim fishers, who tend to be relatively recent entrants into the fisheries but have closer connections to local power brokers and access to higher efficiency harvesting gear, are usually given priority and allocated the most productive locations (Kremer, 1994b).

Apart from encouraging the use of more intensive harvesting technology, this system has led to overfishing. Leaseholders, in a bid to recover the lease payment and bribes they may have paid along the way, usually sublease or allow harvesting by as many fishers as are willing to pay user fees (Ullah, 1985; Naqi, 1989; Aguero and Ahmed, 1990; McGregor, 1995). While fishing regulations are incorporated in the lease agreement in an effort to sustain productivity, in practice the lessee is seldom constrained by them. Thus, in effect, the policy has discouraged long-term investment in fisheries management and has reinforced unequal power structures that have kept poor fishers dependent on a predatory class for continued use of a resource so vital to their subsistence (Figure 7.3).

In an attempt to arrest fisheries degradation and correct the gross inequity in their access, the government, acting on a strong recommendation from the DOF, initiated the New Fisheries Management Policy (NFMP) on an experimental basis in 300 water bodies in 1986 (MOFL, 1986). Through a system of annual

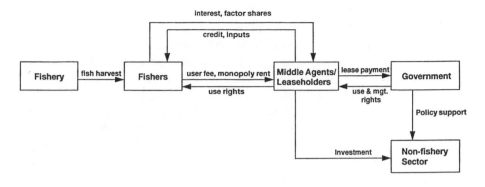

Figure 7.3 Rights and benefit distribution under short-term fisheries leasing policy. (Source: Capistrano et al., 1994.)

gear-specific licensing, access and withdrawal rights under the NFMP were reserved for "genuine" fishers (i.e., those deriving 90% of their total household income from fishing). In return, licensees were expected to abide by and help enforce fishing rules and regulations. Gear-specific licensing was intended to ease the pressure on fisheries by regulating harvesting. Limiting use rights to bona fide fishers was meant to ensure that they get the portion of fishing income supposedly being siphoned off by rent-seeking agents and middlemen.

Essentially, the NFMP was Bangladesh's aquatic equivalent of land reform. It aimed to (1) improve and sustain open water fisheries production, (2) provide traditional and full-time fishers a greater share of fishing income, and, (3) encourage fisheries conservation. Management authority over NFMP-designated water bodies was transferred from the MOL to the MOFL/DOF for the duration of the experiment. However, many of these water bodies remained under *de facto* MOL control and continued to be leased out despite the existence of the NFMP (Naqi, 1989).

Experiments in Management

To test management approaches under the NFMP, in 1987 the Improved Management of Open Water Fisheries Project (IMOF) was jointly undertaken by the DOF, the International Center for Living Aquatic Resources Management (ICLARM) (a Manila-based international research organization), and the Ford Foundation (a private funding and development organization). The IMOF project involved biological research as well as socio-economic monitoring and policy studies and was implemented in two phases (DOF, 1986).

First Phase: IMOF I

During the first phase (1987–89), the DOF took a direct role in identifying "genuine" fishers, issuing licenses, and enforcing fishing regulations in nine water bodies representative of three types of fisheries: (1) flowing rivers, (2) dead rivers and oxbow lakes (baors), and (3) deeper portions of natural floodplain depressions (beels). In each type of fishery, four management approaches were tested: (1) intensive DOF management with additional manpower, infrastructure, and credit provided to DOF-licensed fishers; (2) DOF licensing but no other inputs provided; (3) DOF licensing through NGOs; and (4) usual leasing through the MOL, which served as the control (Rahman, 1989). In the treatment sites, the DOF collected the license fees for the MOL, which were to be paid at the start of the fishing season. The DOF also provided technical management assistance to fishers, although management mostly took the form of limiting fishing effort through regulatory enforcement.

Results indicated that, compared to leasing, licensing appeared to encourage greater fisheries protection and adherence to fishing regulations. This was observed particularly in the beels and baors, which, compared to flowing rivers, have more readily definable boundaries, are easier to manage, and offer participants greater assurance of capturing the conservation benefits of regulatory compliance. Local fisheries management committees, which began to form as a result of the project, helped minimize conflicts and ensure greater equity in the distribution of licenses for specified gear and fishing grounds (Aguero and Ahmed, 1990).

Results show that pure profits as well as returns to labor and capital in flowing rivers under licensing were higher compared to leasing (Table 7.1). However, in beels and baors, licensing yielded unambiguously higher returns over leasing only when credit and other complementary inputs were provided by the DOF (BCAS, 1988). No conclusions can be drawn regarding performance in sites supposed to have been under NGO management because, unfortunately, NGO participation in this phase of the experiment was uneven.

Several important lessons were learned from the initial phase. In a classic oversight characteristic of many early land-reform-type programs, inadequate provision was made for fishers' access to credit and other complementary fishing inputs, even in the intensively managed DOF sites. The (mistaken) assumption was that having been "freed" from middlemen and assured license for the year, the fishers would want to, and be able to, secure credit for necessary inputs from banks and other institutional lenders. The results indicate that without incorporating workable mechanisms to provide credit as well as services performed by moneylenders and middlemen, a shift from leasing to licensing alone would have limited benefits for the fishers (Naqi, 1989).

Table 7.1   Return to Capital, Labor, and Profits (in Bangladeshi Taka) Under Alternative Forms of Management in Three Types of Water Bodies

Type of management	Return to capital (%)	Return to labor (%)	Pure profit (BD taka)
Flowing rivers			
DOF licensing 1 (credit, other inputs provided)	36	189	6,529
DOF licensing 2 (credit, other inputs not provided)	64	242	6,771
Leasing by MOL	20	70	3,009
Floodplain depression (beels)			
DOF licensing 1 (credit, other inputs provided)	408	1,052	47,137
DOF licensing 2 (credit, other inputs not provided)	60	236	6,052
Leasing by MOL	131	286	9,791
Dead rivers (baors)			
DOF licensing 1 (credit, other inputs provided)	44	157	8,944
DOF licensing 2 (credit, other inputs not provided)	57	107	561
Leasing by MOL	26	42	5,957

Source: BCAS Annual Report (1988).

Although licenses were issued in the name of bona fide fishers, the money-lenders and middlemen, who largely opposed and resisted the project, continued to provide fishers credit for license fees and fishing inputs but under more stringent terms than before. To a large extent, production relations and distribution of output in the sample fisheries proceeded as usual. Moreover, in the absence of alternative employment or additional income sources for fishers, and despite limited conservation attempts in certain water bodies, fishing pressure continued unabated.

The schizophrenic role assigned to the DOF (providing technical assistance and extension while collecting fees and enforcing regulations) made effective management extremely difficult and at times downright impossible. Lacking magisterial and police power, the DOF has had to depend on the cooperation of the local police and the deputy commissioner in dealing with violators of fisheries regulations. The DOF's conflicting roles and limited authority created a context in which trust, the basis of management agreements, remained elusive. Not surprisingly, yearly licenses with the possibility, but no guarantee, of indefinite renewal did not give poor fishers the security of resource use rights, which the NFMP had hoped to provide.

Second Phase: IMOF II

A subsequent phase of the project was designed in 1989, but bureaucratic delays pushed back implementation to mid-1992 (DOF, 1989). In the second phase (1992–95), the DOF was supposed to collaborate with four of the country's largest and leading NGOs: the Bangladesh Rural Advancement Committee (BRAC), Proshika Manobik Unnayan Kendra (Proshika), Caritas-Bangladesh (Caritas), and Friends in Village Development, Bangladesh. To help ensure that licenses were given only to the intended beneficiaries, NGOs were supposed to identify "genuine" fishers for licensing by the DOF. As partners in the action-research, the NGOs were expected to perform several roles: (1) help encourage greater participation of fishers in management, (2) utilize their own financial and organizational resources to provide fishers credit and input support, and (3) use their experience in social mobilization and human capacity building to create alternative or supplementary income opportunities for fishers in the project sites (Figure 7.4). DOF was supposed to provide technical fisheries management support and administrative facilitation. The second phase turned out to be an ambitious experiment in government–NGO–fishers co-management, with few rivals in the world (Berkes, 1994).

Coverage was expanded from 9 to 19 sites to include a wider range of fishing environments. Socio-economic monitoring was conducted in all sites. Sampling of catch composition for different gear types was conducted in four water bodies representative of flowing rivers (Narisa-Padma River), deeper portions of flood-plain depressions called beels (Hamil Beel), shallower portions of floodplain

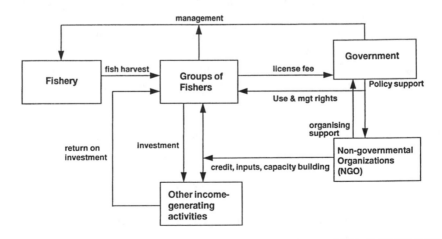

Figure 7.4 Rights allocation and benefit distribution under Bangladesh's 1986 new fisheries management policy of licensing full-time fishers. (Source: Capistrano et al., 1994.)

depressions called haors (Kanglar Haor), and dead rivers or oxbow lakes called baors (Jalkeshab Rai Baor).

Project implementation was complicated by staff turnover in ICLARM and in some of the NGOs, delays in fielding the required technical expertise, and government red tape, all of which greatly affected the quality of research. Furthermore, a drought in 1994–95 biased the data, making it impossible to draw firm conclusions. Additional data need to be collected over a longer term to determine the real impact of controlling harvest through licensing in combination with several types of development interventions. Nevertheless, this phase offers useful lessons which already provide a basis for further refining management approaches in similar environments.

Results indicated an improvement in fishers' income and socio-economic conditions because of the project (Hossain, 1994). However, this was due largely to the effect of NGOs' poverty alleviation activities rather than to increases in the fisheries' productivity as such, although there was some indication of productivity improvement in the sample haor and flowing river fisheries after the first year (Table 7.2). Drought in the succeeding years caused shrinkage in water areas and dried up the baor. Fish catch, the basis for productivity measurement, decreased from 1994 to 1995 in both the sample haor and flowing river (Table 7.2). In Hamil beel, where an increased catch was observed, the increase was due to stocking with carp and a voluntary seasonal closure of the fishery, management measures undertaken collectively by project participants (Ahmed et al., 1992). That licensees had a certain degree of control over access to the beel, and hence some assurance of enjoying the benefits, made it worth their while to invest in these management interventions. This control was weak or lacking in the other sites due to either the nature of the water bodies or the state of organization of the fishers.

Table 7.2 Average Monthly Catch (in kg) from IMOF II Sample Sites

	Year		
Type of water body	*1993*	*1994*	*1995*
Flowing river (Narisha Padma)	1797	2320	1359
Deep floodplain depression (beel: Hamil Beel)	na	486	756
Floodplain depression (haor: Kanglar Haor)	4114	5436	4230
Dead river (baor: Jalkeshab Rai)	1006	981	na

NA = not available.

Source: Hossain (1994).

The preparation of lists of "genuine" fishers took up a major portion of the NGOs' time. Getting the lists approved by the appropriate local government agencies proved most difficult and controversial. A major issue was the exclusion of equally needy subsistence fishers from water bodies regarded as common property, creating conflicts between licensees and excluded "non-genuine" fishers. Likewise, the exclusion of local power brokers from the project invited hostility, non-cooperation, and even sabotage. Non-compliance of unauthorized fishers made it more difficult to enforce regulations even among the licensees— in other words, free-riding by non-licensees encouraged free riding among licensed fishers (Naqi, 1989; Berkes, 1994).

Anxiety over the terms of yearly licensing and up-front fee payment remained. Strong recommendations were made for longer term licenses and the payment of license fees in installments. Because they impose service charges and social development conditions for granting credit (for example, participation in health and literacy programs), NGOs were regarded by some as being no different than traditional moneylenders. There were also tensions in operationalizing DOF–NGO collaboration. The greatest challenge was reconciling their differing priorities, points of view, and modes of operation, particularly at the field level, where interactions tended to be somewhat competitive.

The results suggest that where the characteristics of the resource and the degree of social organization provide user-managers reasonable assurance that they can capture a major portion of the benefits from management, approaches based on targeted empowerment and poverty alleviation can promote resource conservation (Ahmed et al., 1992; Capistrano, 1996). In this second phase of the collaborative project, the NFMP's definition of "genuine" fishers obviously needed to be clarified and its narrow focus on this group of users re-examined. While continuing to be explicit about the intended primary beneficiaries, for practical purposes, there was a need to expand beyond a class of fishers to the larger social units in which they are embedded (i.e., to take account of other stakeholders and of their varying levels of influence on the success or failure of any management intervention) (Capistrano et al., 1994). Ultimately, management approaches framed through negotiations among the water bodies' various users and stakeholders, including those who may not be fishers but whose activities impact on the fisheries, can be the only viable long-term solution to reversing the resource's continuing degradation.

Community-Based Fisheries Management

The DOF, ICLARM, and several NGOs, including BRAC, Proshika and Caritas, and the Ford Foundation, are again collaborating on a new action-research project on community-based fisheries management now in its early stages of implementation (DOF, 1995). This project will encourage flexible and pragmatic management approaches involving communities (groups of users and stakeholders) that simultaneously address issues of poverty alleviation, social equity, and fisheries conservation whether under leasing or licensing arrangement. It will test alternative models of partnership among the government, NGOs, and communities and examine their contribution to several processes: (1) improving the socio-economic conditions of several categories of fishers, (2) reducing pressure on the fisheries, (3) encouraging community participation, and (4) creating more equitable local institutions. It will build on working relationships forged among agencies and incorporate lessons learned from the IMOF experience, mindful of the influence of power structures in determining outcomes.

This project is one of several supported by the Ford Foundation in Bangladesh under its Community-Based Fisheries Management Initiative started in 1993. The initiative aims to promote means that would empower the rural poor to access, sustainably manage, and derive benefits from common property fisheries. To this end, it funds grass-roots experimentation, advocacy, and policy analysis and formulation, as well as programs to amplify grass-roots voices and strengthen accountability at various levels of policy making and implementation (Capistrano, 1994). Among others, the initiative currently underwrites experiments in community-based wetlands restoration, legal research on the country's regulatory framework for fisheries, a forum for NGO advocacy on fisheries-related development issues, and studies on the human ecology and dynamics of Bangladesh's fisheries. The new project and these various other efforts would, hopefully, help foster more informed debate on appropriate policies and approaches for fisheries management in Bangladesh and in other countries faced with similar challenges.

Continuing Challenges

Indeed, the challenges are many. A recent challenge is the confusion over a September 1995 policy authored by the MOL practically abolishing the NFMP and strongly reaffirming a leasing policy without consultation with the DOF (MOL, 1995). A system of licensing will be in place only in flowing rivers, where annual renewable licenses will be issued by the district deputy commissioner, who will also determine the license fee schedule based on the profitability of the boats and gear used. In all other water bodies, revenue-based competitive leasing will remain in force. Exceptions are water bodies currently being managed by the DOF under various existing fisheries development projects. Upon completion of the projects, these too will be leased out.

Although the protection of poor fishing communities is used to justify the policy, it really is a means of perpetuating and strengthening local power structures now controlling fisheries allocation and access. The issue of access and control becomes more crucial as the stakes become higher. For larger commercial fishers and their agents, government stocking programs and the possibility of applying aquaculture technology to open waters promise greater returns than ever before. Their desire for larger shares of potential profits reinforces a tendency toward greater exclusion of other resource users. For subsistence fishers and poor resource users, continued access to the water bodies is becoming even more important as population and unemployment increase while fisheries continue to dwindle.

Conflicts over water bodies are bound to escalate unless creative and pragmatic solutions are found. With their grounding in research and links to policy advocacy and formulation, hopefully, community-based efforts toward local fisheries management now ongoing can help chart the direction of possible solutions.

References

Aguero, M. and Ahmed, M. (1990) Economic rationalization of fisheries exploitation through management: experiences from the open water inland fisheries management in Bangladesh. In: *Proceedings of the Second Asian Fisheries Forum*. R. Hirano and I. Hanyu, Eds. Manila, Philippines, 991 pp.

Aguero, M., Huq, S., Rahman, A.K.A., and Ahmed, M., Eds. (1989) *Inland Fisheries Management in Bangladesh*. Department of Fisheries, Dhaka, Bangladesh; Bangladesh Center for Advanced Studies, Dhaka, Bangladesh; and International Center for Living Aquatic Resources Management, Manila, Philippines, 149 pp.

Ahmed, A.T.A. (1995a) Aquatic biodiversity: Bangladesh scenario, paper presented at the Fourth Asian Fisheries Forum Meeting, Beijing, China.

Ahmed, A.T.A. (1995b) Impacts of other sectoral development on the inland capture fisheries of Bangladesh, paper presented at the Fourth Asian Fisheries Forum Meeting, Beijing, China.

Ahmed, M.N. and Hossain, M.M. (1995) Legal, regulatory and institutional framework for fisheries and fishing community development and management, paper presented at the National Workshop on Fisheries Resources Development and Management, Dhaka, Bangladesh.

Ahmed, M., Capistrano, D., and Hossain, M. (1992) Redirecting benefits to genuine fishers: Bangladesh's new fisheries management policy. *Naga ICLARM Quarterly (Bangladesh)*, 15(4): 31–34.

Ahmed, M., Capistrano, A.D., and Hossain, M. (1995) Fisheries co-management in Bangladesh: experiences with GO-NGO-fishers partnership models, paper presented at the Fifth Common Property Conference of the International Association for the Study of Common Property, Bodoe, Norway, May 24–28.

Ali, M.H., Miah, N.I., and Ahmed, N.U. (1993) *Experiences in Deepwater Rice–Fish Culture.* Bangladesh Rice Research Institute, Gazipur, Bangladesh.

Ali, M.L. and Fisher, K. (1995) Potential, constraints and strategy for conservation and management of inland open water fisheries in Bangladesh, paper presented at the National Workshop on Fisheries Resource Development and Management. MOFL, ODA, and FAO, Dhaka, Bangladesh.

Ali, M.Y. (1991) *Towards Sustainable Development of Fisheries Resources of Bangladesh.* National Conservation Strategy of Bangladesh, International Union for Conservation of Nature and Natural Resources, and Bangladesh Agricultural Research Council, Dhaka, Bangladesh, 96 pp.

ANGOC (1989) *People's Participation and Environmentally Sustainable Development.* Asian NGO Coalition for Agrarian Reform and Rural Development, Manila, Philippines.

Barbier, E.B. (1987) The concept of sustainable economic development. *Environmental Conservation,* 14(2): 101–110.

Bardham, P. (1993) Symposium on management of local commons. *Journal of Economic Perspectives,* 7(4): 87–92.

BCAS (1988) Annual Report. Experiments in New Approaches to the Management of Inland Open Water Fisheries of Bangladesh. Bangladesh Center for Advanced Studies, Dhaka, Bangladesh.

Berkes, F. (1989) *Common Property Resources: Ecology and Community-Based Sustainable Development.* Belhaven Press, London, U.K., 302 pp.

Berkes, F. (1994) Improved Management of Openwater Fisheries (IMOF) Project Evaluation Mission Report. International Center for Living Aquatic Resources Management, Manila, Philippines.

Boyce, J.K. (1994) Inequality as a cause of environmental degradation. *Ecological Economics,* 11: 169–178.

Bromley, D.W. (1991) *Environment and Economy: Property Rights and Public Policy.* Basil Blackwell, Oxford, U.K., 247 pp.

Capistrano, A.D. (1994) *Community-Based Fisheries Management in Bangladesh: A Program Strategy.* The Ford Foundation, Dhaka, Bangladesh, 20 pp.

Capistrano, A.D. (1996) Participatory resource management in the context of growth-inducing, self-empowering poverty alleviation. In: *Competition and Conflict in Asian Agricultural Resource Management: Issues, Options and Analytical Paradigms.* P. Pingali and T. Paris, Eds. International Rice Research Institute, Laguna, Philippines.

Capistrano, A.D., Ahmed, M., and Hossain, M. (1994) Ecological economic and common property issues in Bangladesh's openwater and floodplain fisheries, paper presented at the Third Biennial Meeting of the International Society for Ecological Economics, San José, Costa Rica, Oct. 24–28, 10 pp.

Charles, A.T. (1994) Towards sustainability: the fishery experience. *Ecological Economics,* 11: 201–211.

Coase, R.H. (1960) The problem of social cost. *Journal of Law and Economics,* 3: 1–44.

Daly, H.E. and Cobb, J.B. (1989) *For the Common Good*. Beacon Press, Boston, MA.

Dietz, F.J. and Straaten, J. (1992) Sustainable development and the necessary integration of ecological insights into economic theory. In: *Sustainability and Environmental Policy: Restraints and Advances*. F.J. Dietz, U.E. Simonis, and J. van der Straaten, Eds. Edition Sigma, Berlin, Germany, 295 pp.

DOF (1986) Technical Assistance Project Proforma for Improved Management of Openwater Fisheries. Government of Bangladesh Ministry of Fisheries and Livestock, Dhaka, Bangladesh.

DOF (1989) Technical Assistance Project Proforma for Improved Management of Openwater Fisheries (Phase II). Government of Bangladesh Ministry of Fisheries and Livestock, Dhaka, Bangladesh.

DOF (1995) Technical Assistance Project Proforma for Community-Based Inland Openwater Fisheries Management and Development Project. Government of Bangladesh Ministry of Fisheries and Livestock, Dhaka, Bangladesh.

Dyer, C.L. and McGoodwin, J.R., Eds. (1994) *Folk Management in the World's Fisheries: Lessons for Modern Fisheries Management*. University Press of Colorado, Niwot, CO.

FAP–17 (Flood Action Plan 17) (1993) Nature and Extent of NGOs' Participation in Fisheries Resource Development in Bangladesh. Interim Report prepared for the Government of Bangladesh. Dhaka, Bangladesh.

Farooque, M. (1989) Laws on wetlands in Bangladesh: a complex legal regime. In: *Freshwater Wetlands in Bangladesh: Issues and Approaches for Management*. A. Nishat, Z. Hussain, M.K. Roy, and A. Karim, Eds. IUCN, Gland, Switzerland, pp. 231–237.

FRSS (1994) Bangladesh Fisheries Statistics. Department of Fisheries, Ministry of Fisheries and Livestock, Dhaka, Bangladesh.

Gadgil, M. and Berkes, F. (1991) Traditional resource management systems. *Resource Management and Optimization*, 8(3–4): 127–141.

Gibbs, J.N. and Bromley, D.W. (1989) Institutional arrangements for management of rural resources: common property regimes. In: *Common Property Resources; Ecology and Community-Based Sustainable Development*. F. Berkes, Ed. Belhaven Press, London, U.K., 302 pp.

Goulet, D. (1989) Participation in development: new avenues. *World Development*, 17(2): 165–178.

Hardin, G. (1968) The tragedy of the commons. *Science*, 162: 1243–1248.

Hossain, M. (1994) Annual Report of Improved Management of Open Water Fisheries Phase II. Government of Bangladesh Department of Fisheries, Dhaka, Bangladesh.

ISPAN (1993) Flood Control and Nutritional Consequences of Biodiversity of Fisheries. Environmental Study (FAP 16) prepared for the Flood Plan Coordination Organization of the Ministry of Irrigation Water Development and Flood Control. Irrigation Support Project for Asia and the Near East, Arlington, VA.

Jazairy, I., Alamgir, M., and Panuccio, T. (1992) *The State of World Rural Poverty*. International Fund for Agricultural Development, IT Publications, London, U.K., 514 pp.

Khan, A.A. (1993) Freshwater wetlands in Bangladesh: opportunities and options. In: *Freshwater Wetlands in Bangladesh: Issues and Approaches for Management.* A. Nishat, Z. Hussain, Z., M.K. Roy, and A. Karim, Eds. IUCN, Gland, Switzerland, pp. 1–8.

Kremer, A. (1994a) The Impact Upon Income Distribution of an Intensification of Inland Fisheries in Developing Countries: Three Theorems. Occasional Paper 01/94. Center for Development Studies, University of Bath, U.K.

Kremer, A. (1994b) *Equity in the Fishery: A Floodplain in N.E. Bangladesh.* Bath University Centre for Development Studies, U.K., and Prince of Songkla University Coastal Resources Institute, Thailand.

Lovell, C. H. (1992) *Breaking the Cycle of Poverty: The BRAC Strategy.* Kumarian Press, West Hartford, CT.

McGregor, J.A. (1995) The assessment of policy and management options in inland capture fisheries: summary guidelines, presented at the seminars on Fisheries Research Dissemination, University of Bath, U.K., and Proshika Manobik Unnayan Kendra, Bangladesh.

Minkin, S.F. and Boyce, J.K. (1994) Net losses: "development" drains the fisheries of Bangladesh. *Amicus Journal,* 16(3): 36–40.

MOFL (1986) Outline of a New Management Policy for Public Water Bodies in Inland Fisheries. Government of Bangladesh Ministry of Fisheries and Livestock, Dhaka, Bangladesh.

MOL (1995) Notice. Section—7. Miscellaneous—11/95/576. 20 Bhadra 1402/ 4 September 1995. Ministry of Land, Dhaka, Bangladesh.

Naqi, S.A. (1989) Licensing versus leasing system for fishing access. In: *Inland Fisheries Management in Bangladesh.* M. Aguero, S. Huq, A.K.A. Rahman, and M. Ahmed, Eds. Department of Fisheries, Dhaka, Bangladesh; Bangladesh Center for Advanced Studies, Dhaka, Bangladesh; and International Center for Living Aquatic Resources Management, Manila, Philippines, pp. 83–92.

Naqi, S.A., Ali, M.Y., Sadeque, S.Z., and Khan, T.A. (1994) Study on Increased Intervention by NGOs in the Third Fisheries Project Floodplains: Final Report. Bangladesh Center for Advanced Studies, Dhaka, Bangladesh.

Nishat, A., Hussain, Z., Roy, M.K., and Karim, A., Eds. (1993) *Freshwater Wetlands in Bangladesh: Issues and Approaches for Management.* IUCN, Gland, Switzerland, 283 pp.

Oodit, D. and Simonis, U. (1992) Poverty and sustainable development. In: *Sustainability and Environmental Policy: Restraints and Advances.* F. Dietz, U. Simonis, J. van der Straaten, Eds. Edition Sigma, Berlin, Germany.

Ostrom, E. (1990) *Governing the Commons: The Evolution of Institutions for Collective Action.* Cambridge University Press, New York, NY.

Ostrom, E. and Gardner, R. (1993) Coping with asymmetries in the commons: self-governing irrigation systems can work. *Journal of Economic Perspectives,* 7(4): 93–112.

Pearce, D.W. (1988) The sustainable use of natural resources in developing countries. In: *Sustainable Environmental Management: Principles and Practice.* R.K. Turner, Ed. Belhaven Press, London, U.K.

Pezzey, J. (1992) Sustainable Development Concepts: An Economic Analysis. World Bank Environment Paper Number 2. The World Bank, Washington, D.C.

Rahman, A.K.A. (1989) The new management policy of open water fisheries in Bangladesh under experimental monitoring and evaluation. In: *Inland Fisheries Management in Bangladesh.* M. Aguero, S. Huq, A.K.A. Rahman, and M. Ahmed, Eds. Department of Fisheries, Dhaka, Bangladesh; Bangladesh Center for Advanced Studies, Dhaka, Bangladesh; and International Center for Living Aquatic Resources Management, Manila, Philippines, pp. 14–23.

Rahman, A.K.A. (1993) Wetlands and fisheries. In: *Freshwater Wetlands in Bangladesh: Issues and Approaches for Management.* A. Nishat, Z. Hussain, M.K. Roy, and A. Karim, Eds. IUCN, Gland, Switzerland, pp. 147–161.

Schlager E. and Ostrom, E. (1992) Property rights regimes and natural resources: a conceptual analysis. *Land Economics,* 68(3): 249–262.

Seabright, P. (1993) Managing local commons: theoretical issues in incentive design. *Journal of Economic Perspectives,* 7(4): 113–134.

Siddiqi, K. (1989) Licensing versus leasing system for government-owned fisheries (jalmahals) in Bangladesh. In: *Inland Fisheries Management in Bangladesh.* M. Aguero, S. Huq, A.K.A. Rahman, and M. Ahmed, Eds. Department of Fisheries, Dhaka, Bangladesh; Bangladesh Center for Advanced Studies, Dhaka, Bangladesh; and International Center for Living Aquatic Resources Management, Manila, Philippines, pp. 73–82.

TSS–1 Fishery Sector Programming Mission to Bangladesh (1992) National Fishery Development Program. Ministry of Fisheries and Livestock, Dhaka and Food and Agriculture Organization, Rome, Italy.

Ullah, M. (1985) Fishing rights, production relations and profitability: a case study of Jamuna River fishermen in Bangladesh. In: *Small-Scale Fisheries in Asia: Socio-Economic Analysis and Policy.* T. Panayotou, Ed. International Development Research Center, Ottawa, Canada.

Watanabe, T. (1993) *The Ponds and the Poor: The Story of Grameen Bank's Initiative.* Grameen Bank, Dhaka, Bangladesh, 97 pp.

WCED (1987) *Our Common Future.* World Commission on Environment and Development and Oxford University Press, Oxford, U.K., 400 pp.

Wood, G.D. (1984) Provision of irrigation assets by the landless: an approach to agrarian reform in Bangladesh. *Agricultural Administration,* 17(2): 55–80.

Wood, G. and Gregory, R. (1990) Off the page and into the pond, paper for the Workshop on Production, Training and Extension Strategies of the Hatchery Development Project, Dinajpur, Bangladesh.

Wood, G.D. and Palmer-Jones, R. (1991) *The Water Sellers.* IT Publications, London.

World Bank (1987) Environment, Growth and Development. Development Committee Pamphlet 17. World Bank, Washington, D.C.

Wildlife Use for Economic Gain: The Potential for Wildlife to Contribute to Development in Namibia

8

Caroline Ashley and Jonathan Barnes
Directorate of Environmental Affairs, Ministry of Environment
and Tourism, Windhoek, Namibia

Abstract

The degree to which wildlife utilization can contribute to sustainable and equitable development in Namibia is explored in this chapter. In the past, wildlife was not regarded as a motor for economic growth or rural development. However, experience on commercial farms supplemented by economic research shows that the economic benefits from wildlife utilization could more than double over 10 to 20 years. Local incomes in some historically marginalized rural areas could increase severalfold. Key to this development is the establishment of appropriate property rights for residents, plus opportunities to realize high returns from using wildlife within these property rights. In the commercial farming sector, many of these conditions are in place and wildlife use has already expanded. In the communal areas, the appropriate policy framework is only just being established, and much needs to be done to ensure that benefits to residents from wildlife are optimized. Analysis shows that development of community rights, support for communities and local enterprises, development of wildlife as

1-57444-077-2/97/$0.00+$.50
© 1997 by CRC Press LLC

a complement to agriculture, minimization of land use trade-offs, and inclusion of non-consumptive wildlife uses are important components of a strategy for this.

Introduction

Namibia has a rich and rare environmental endowment, such as the ancient welwitschia plant (*Welwitchia mirabilis*), around 700 species of endemic beetles, and elephants (*Loxodonta africana*) adapted to desert conditions. Wetter parts of the country support the more typical African game, including the "big five"— elephant, black rhinoceros (*Diceros bicornis*), lion (*Panthera leo*), leopard (*Panthera pardus*), and buffalo (*Syncerus caffer*). Spectacular scenery includes rolling sand dunes of the desert, the wilderness of Kaokoland, and lush rivers and floodplains of Caprivi. The network of protected areas includes the world-famous Etosha National Park.

These environmental assets have long been important to conservationists from around the world and are increasingly important to tourists, but have not in the past been important to the majority of Namibians. However, there is now growing evidence that Namibia's environmental wealth can make a substantial contribution to the country's post-apartheid development through the principle of sustainable utilization. In a newly independent nation, where land tenure, income, and skills are still highly skewed, where cattle is a cultural and economic mainstay for many, where natural resources are at risk of degradation, and where more equitable and diversified development are national goals, wildlife utilization can bring profits, growth, equity, and sustainability.

This chapter outlines the current and potential economic contribution of Namibia's wildlife resources and highlights some of the steps that must still be taken if this development potential is to be realized. The first half of the chapter explores the contribution of wildlife and tourism to the national economy. The second half focuses on the contribution to local incomes and development in the poorer regions, the "communal areas." Throughout the chapter, values are given in Namibian dollars (N$), where N$1.00 = 1 South African rand and N$3.65 = US$1 (1995 exchange rate).

Background and Context

Namibia is a country of 1.6 million people (National Planning Commission, 1994) and 824,000 km^2 (Brown, 1994), located in the southwestern tip of Africa. It comprises a narrow western coastal plain, from which the land rises 1000 m

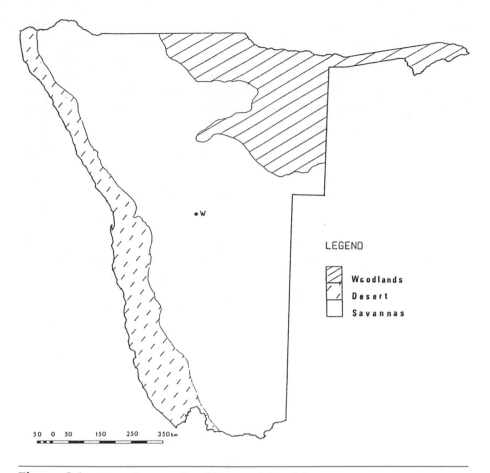

Figure 8.1   Natural vegetation biomes of Namibia.

or so to an extensive interior plateau. Namibia has the driest climate of any country south of the Sahel, and much of the country is desert or semi-desert, with no perennial rivers between its northern and southern borders.

Corresponding primarily with rainfall, but also with soil characteristics, there are three major vegetation zones: desert occupies the western coastal plain and the south, savannah occupies the central and north-central plateau, and woodland occupies the wetter northeast (Figure 8.1). The distribution of large mammals and other so-called "charismatic macrofauna" of interest here corresponds to these zones. A few arid-adapted species are found in the desert, for example, gemsbok (*Oryx gazella*) and springbok (*Antidorcas marsupialis*). A slightly more diverse plains game community is found in the central savanna, of which there is a large number of endemic species, such as the Hartmanns mountain zebra (*Equus zebra hartmannae*). A relatively rich fauna and the highest wildlife biomass occupy the northeast, including such central African elements as the lechwe (*Kobus leche*) and sitatunga (*Tragelaphus spekei*) (see Joubert and Mostert, 1975). Apart from the leopard, which is widespread, the rest of the so-called "big five" wildlife species tend to be concentrated in the northern state lands, including both protected areas and communal land.

Until 1990, Namibia was occupied by South Africa. Consequences of apartheid rule still pervade, as in the grossly unequal distribution of income and land. The country is divided into commercial farmland (43%, mainly in the savanna and semi-desert areas of the south and center) and communal land (former "homelands," 40%, largely in the north), as shown in Figure 8.2. On both, livestock farming predominates, as most of the country is too dry for arable farming, but in all other respects the differences are extreme. Commercial land is privately owned by approximately 4600 mainly white farmers (less than 1% of the population). These private farms average over 7000 ha in size. Extensive livestock ranching is mostly of cattle in the center/north and sheep in the arid south, for commercial sale and export. The majority of Namibians live in communal areas, where the land is state owned and farmers have only usufruct rights. Crops are produced on small individually allocated plots of a few hectares in limited areas of the north where soils are suitable and water available, but grazing is done in commonly managed or open-access areas. For most communal farmers, livestock serve many purposes, providing milk, draught power, meat, manure, a mark of status, a store of wealth, and other social functions. Veterinary barriers prevent movement of livestock and unprocessed livestock products from most northern communal areas to the south. Agricultural incomes are so low and variable that cash remittances and pensions are essential supplements for most families, and 17% of rural households regard these as their main source of income (Central Statistical Office, 1995a).

Of the remaining state-owned land, some 13% is covered by 14 protected areas, and 2% is reserved for diamond mining (Brown, 1994). Mining is by far the largest economic sector in terms of contribution to gross national product. Another large component of the economy is marine fishing based on the productive, cold water upwellings of the Benguela current, but commercial livestock ranching (8% of gross domestic product [GDP] in 1994) and communal subsistence livestock (largely unmeasured) provide the livelihood of the vast majority and form the main land use in the country. The "land question" remains unresolved: there is pressure for redistribution, but much of the commercial farmland is unsuitable for uses other than livestock keeping, with between 10 and 25 ha needed per large stock unit.

GDP was N$10.4 billion in 1994 (US$2.9 billion) (Central Statistical Office,

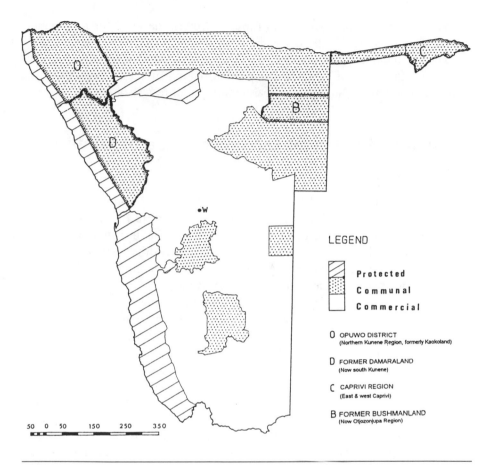

Figure 8.2 Land tenure in Namibia. O = Opuwo District (Northern Kunene Region, formerly Kaokoland), D = former Damaraland (now South Kunene), C = Caprivi region (East and West Caprivi), B = former Bushmanland (now Otjozonjupa region).

1995c). However, the relatively high average per-capita income (US$1865) masks a sharply dualistic economy. Average annual per-capita income among the top 10% of households is about N$17,500 compared to N$1500 in the rest of the population. The top 1% have a total annual household income that exceeds the total income of the bottom 50% (Central Statistical Office, 1995b).

Only a third of the active population is employed in the formal sector (Government of the Republic of Namibia, 1995). Unemployment is estimated at around 20%, with a further 40% estimated to be underemployed (Central Statistical Office, 1995b). As the population is growing faster than the economy (Government of the Republic of Namibia, 1995), and few formal sector jobs are found in the more populous north (Tapscott, 1992), the need for more labor-intensive and geographically dispersed growth is urgent.

The harsh climate, unequal access to land and income, the tradition of livestock, and priorities of a newly independent nation affect all aspects of the political economy in Namibia and particularly wildlife utilization.

Current and Potential Economic Value of Wildlife

Wildlife Uses

Wildlife occurs in varying densities on nearly all land in Namibia. The legislative and policy framework that permits the use of wildlife for economic commercial gain reflects the legacy of the apartheid era in that private (commercial) landholders have custodial rights to manage and use wildlife on their land while those on communal lands do not. A new policy has been developed, and legislative changes are being enacted to make it possible for communal landholders to acquire common property rights over wildlife resources on their lands. The delegation of control over the wildlife resources from central government to local communities in communal land will be possible through the development of wildlife and natural resource "conservancies."*

The use of wildlife in Namibia has involved non-consumptive tourism, consumptive tourism (recreational hunting and fishing), and consumptive use for meat, skins, and other products (Joubert, 1974; Yaron et al., 1994). Overall, non-consumptive tourism, based on viewing wildlife and wilderness, dominates, but there are important differences between protected, commercial, and communal areas. Wildlife viewing activities are centered around the protected areas, particularly Etosha National Park and the Sossusvlei sand dunes. However, the fastest growth in tourism is now occurring outside the parks, with a mushrooming of guest farms and lodges on commercial land and lodges and specialized tours in communal areas.

Consumptive uses of wildlife have tended to be concentrated on commercial farmland, where the majority of people use game for their own family, friends, and workers (Yaron et al., 1994). Recreational hunting, mainly of plains game for sport rather than for trophies, is a common form of wildlife use on private farms. In addition, over 400 farms are registered as hunting farms to host trophy hunters. Off-take for commercial sale of venison is focused on springbok in the

---

* Conservancies currently occur on private land where farmers group together to manage and use their wildlife. The new policy is to extend this concept to communal land, giving communities common property and custodial rights over wildlife on their land.

south and kudu and gemsbok in the north. Survey returns from commercial farmers to the Ministry of Environment and Tourism indicate that of the 100,000 or so animals shot per year, roughly equal numbers are killed for personal uses, for hunting, and for commercial sale (Yaron et al., 1994).

Consumptive uses of wildlife in northern communal areas are mainly through government-controlled trophy hunting for the "big five" species. Legal local hunting for feasts or annual culls occurs on a small scale.

Estimating the economic value of these wildlife uses is a matter of piecing a jigsaw puzzle together. Some pieces are missing or roughly hewn, but there is sufficient evidence to indicate that the economic benefits of wildlife on commercial land have grown rapidly in the last 20 years; that economic and local benefits on communal land have the potential to multiply; and that the protected areas, by anchoring the tourism industry, are maintaining one of the most important sectors of the Namibian economy.

These jigsaw pieces are presented in the next section, which focuses on the *economic* contribution of wildlife enterprises (i.e., the net contribution to national welfare measured as *net value added to national income*).* This is different from the estimates of *financial* benefit accruing to investors in a specific enterprise or from estimates of local revenue earned by community members, which is outlined in a later section.

Value of Wildlife on Private Land: 20 Years of Growth

On private land, the number of species used for game has increased by 44% over 20 years, while the total number of animals and biomass has increased by 80%, according to questionnaire surveys for 1972 and 1992 analyzed by Barnes and de Jager (1995), the source of figures in Table 8.1 except where otherwise stated. The economic contribution per large stock unit (LSU) equivalent of game averages over N$100/LSU on a typical farm where culling and hunting are supplements to livestock ranching. This average hides extremes between those farmers who make no commercial use of naturally occurring game and those maximizing use through a game lodge devoted to wildlife viewing (where net value added

---

* Net value added to national income, as defined by Gittinger (1982), was derived by subtracting economic costs (including costs of capital) from economic benefits for any given activity. In the process, financial values were converted to economic values, using shadow pricing criteria adopted by the Directorate of Environmental Affairs. The net economic contribution is also a measure of the return to land and government investment because the opportunity cost of land and the economic costs of government expenditures were not deducted. These values were extracted or extrapolated from financial and economic cost–benefit models of resource use activities.

is nearly N$600/LSU). The average net economic contribution (in 1994 prices) of wildlife on private land was N$56 million in 1992 compared to N$31 million in 1972.* This is equivalent to an increase from N$85 to N$157 in net value added per square kilometer.

Although wildlife remains a supplement to, rather than substitute for, livestock on most private land, it is evident that wildlife use has grown faster. As a proportion of the economic value of all private rangeland use, the economic value of wildlife appears to have risen from 5% in 1972 to 11% in 1992.

It is interesting to note that this shift does not seem to be driven by profit maximization on the part of farmers. The effects of sales taxes, rental fees, market wages, and other factors that are paid by farmers but excluded or adapted in the economic model are to make the financial profitability of wildlife use lower than the economic profitability. The investor's financial rate of return is only around 4 to 6% per year for livestock, mixed livestock/wildlife, and pure wildlife. From the national economic perspective, pure wildlife ranching for tourism generates higher returns than mixed livestock/game farming, but not higher financial returns for the investor. Furthermore, on a mixed livestock/ game farm, the income earned per LSU of game is marginally lower than that per LSU of livestock (Department of Environmental Affairs, unpublished data).

These observations suggest that part of the value of game to farmers lies in the diversification of risk and the aesthetic (non-use) benefits not captured in the economic analysis. Diversification is particularly important when farming in such a variable environment as Namibia's, where profit margins are low. The analysis also suggests that policies that are making economically sound wildlife activities financially shaky need to be addressed. To some extent, the relative profitability of wildlife over livestock is likely to increase automatically as trade agreements lower the price of livestock products, while expansion of up-market tourism may increase the returns per LSU of game.

As the profitability of wildlife increases, there will be further incentives to boost wildlife populations. As density and diversity increase, the higher value-added uses of wildlife, such as game lodges and trophy hunting, are in turn likely to continue expanding. The conglomeration of farms into conservancies generates higher returns (both economically and financially) than individual farms, and this trend is also likely to continue. Therefore, a continued expansion of wildlife numbers, and an even faster increase in the total economic contribution of wildlife, is likely. The economic contribution of wildlife on private land to the Namibian economy could effectively double again in the next 10 to 20 years.

---

* Assuming that the use of wildlife and therefore value per LSU was roughly constant in real terms.

Value of Wildlife on Communal Land: Potential to Multiply

In contrast to the commercial areas, the numbers of many wildlife species on communal land appear to have been in decline. Generally, where increases have occurred, they are in areas where community-based conservation initiatives are already in place and they involve larger species, such as black rhinoceros and elephant. Because wildlife on communal land has been classed as state property, there has been little opportunity for residents to benefit from its use and therefore little incentive to conserve wildlife. Furthermore, the wildlife that is present is generally not exploited to its full sustainable potential. In particular, tourism on communal land has developed in an ad hoc way, rather than planned to optimize economic benefits.

Barnes (1995) assessed the economic value of various activities that use wildlife and other non-agricultural natural resources in the four areas of communal land that generally have better wildlife populations and where community-based wildlife conservation projects are in progress: Caprivi region, "former Bushmanland," "former Damaraland," and Opuwo District (see Figure 8.2). Associated protected areas were also included in the study. Together these four areas make up 43% of the communal land in Namibia. Given that livestock is a cultural and economic mainstay in most communal areas, the research focused on wildlife as an addition to agriculture and assumed agricultural activities remained constant. The research gives a picture of the net economic contribution of different activities in 37 zones of the four study areas.* The aggregated results indicate the overall value of wildlife in these four communal areas, while analysis of the components helps to answer key questions such as which areas and which activities have highest potential for increased economic benefits. The results, summarized in Table 8.1, provide the answers to the following questions:

1. *Current and potential economic contribution*—In total, it is estimated that wildlife utilization in the four communal areas currently contributes around N$7.5 million to net national income, ranging from N$6 to N$215 per km^2. If existing resources are used to their sustainable potential, this could more than double to N$16.5 million. Even more, about 2.5 times current value, could be generated with a feasible increase in the resource base.
2. *Comparison between areas*—As Table 8.1 shows, Caprivi generates the highest *absolute* level of economic benefits. However, it is also the

---

* In each zone, the number of current and potential enterprises was estimated and multiplied by the estimated net economic contribution per enterprise. The definition and derivation of net value added to national income are as in the commercial area research above.

Table 8.1　Current and Potential Contribution to National Income of Wildlife Utilization (in Thousands of Namibian Dollars) in Four Study Areas in Communal Land with Associated Protected Areas (Shown in Figure 8.2)[a]

	Caprivi region		Former Bushmanland[b]		Opuwo District		Former Damaraland[c]		Total	
Extent (km²)	18,800		17,877		61,585		58,105		156,367	
	N$	%	N$	%	N$	%	N$	%	N$	%
Current contribution										
Non-consumptive tourism[d]	2,181	53	77	62	1,467	99	1,466	76	5,191	67
Consumptive tourism (hunting, angling)	1,969	47	0	0	0		439	23	2,408	31
Small-scale hunting	9	0.2	48	38	15	1	24	12	119	2
Subtotal	4,159		125		1,482		1,929		7,695	
LESS wildlife damage	110		14		14		30		168	
TOTAL	4,049		112		1,468		1,899		7,528	
Total per km² (N$)	215		6		24		33		48	
Potential contribution										
Non-consumptive tourism	4,851	69	609	58	3,622	10	4,192	86	13,274	80
Consumptive tourism	2,180	31	388	37	0	0	671	14	3,239	20
Small-scale hunting	2	—	60	6	9	—	6	—	77	—
Subtotal	7,033		1,057		3,631		4,869		16,590	
LESS wildlife damage	55		17		14		30		116	
TOTAL	6,978		1,040		3,617		4,839		16,474	
Total per km² (N$)	371		58		58		83		105	

Percentage increase
(current to potential)

Non-consumptive tourism	122	690	147	186	156
Consumptive tourism	11	negligible	0	53	35
Small-scale hunting	–77	25	–66	–75	–35
TOTAL net of wildlife damage[e]	72	828	146	155	119

[a] Adapted from Barnes (1995).

[b] "Former Bushmanland" refers to Tsumkwe District, eastern Otjozondjupa region, north of latitude 22.

[c] "Former Damaraland" refers to the whole of Khorixas District in Kunene region, the western communal land in Erongo region and the West Coast Tourist Recreation Area.

[d] Craft production and marketing are included in non-consumptive tourism, although some items are sold to hunters and local residents.

[e] Damage caused by wildlife to communities (e.g., elephant damage to crops, predation of livestock).

region where utilization is already most developed, so the potential for *expansion* of economic use value ranges from 1.7 times current value in Caprivi to eight times in "former Bushmanland," where commercial wildlife use is currently minimal.

3. ***Values of protected areas***—Communal land adjacent to protected areas has significantly higher current and potential economic value from wildlife use than areas further away. Many of the use values measured in these buffer zones are dependent on the integrity of the associated protected areas. The research also shows that economic benefits generated *inside* the parks and protected areas are currently very low but have potential for enormous (five- and sixfold) increases. Thus, optimal benefits require a change in wildlife utilization inside protected areas as well as on communal land.

4. ***Importance of non-consumptive tourism***—Overall and particularly in the dry but scenic northwestern parts, non-consumptive tourism dominates the current and potential economic use values. The highest returns per square kilometer are derived from non-consumptive tourism. However, as these are only achievable at prime sites, there are large areas of Caprivi (with higher biological productivity and variable potential for wildlife viewing) and Bushmanland (with less scenic attraction) where consumptive wildlife use will be the most viable option.

The evidence indicates that there is considerable latent potential for increasing the contribution that wildlife makes to economic growth in Namibia. On private land, it seems that a policy environment and an array of financial and economic forces have already encouraged an expansion of wildlife use, and this is set to continue. On communal land, economic benefits are currently much smaller, but some areas have potential for severalfold increases. However, much needs to be done to create the right conditions for a similar expansion. For landholders in communal areas to invest land and resources in wildlife conservation, they need a return in benefits from wildlife. Ways in which this can be achieved are discussed in a later section

Wildlife in Protected Areas

The value of wildlife in national parks and game reserves is not easy to assess. Here, the resource, and its use for tourism, has remained under virtually exclusive control of the state. Some of the direct uses occur in the market economy, particularly tourism and the limited capture for live sale, but often not at market prices. Other direct uses, such as research, education, and aesthetic pleasure, cannot be easily valued, while some of the most important values of national parks lie in their indirect benefits and non-use values: maintenance of essential ecological functions and the existence and option value of the biodiversity they preserve. Wildlife is therefore just one component of the assets of a national park. The total annual subsidy for the running of the protected area network (i.e., the total costs of running parks and reserves less receipts from tourists) of around N$30 million per year covers all these benefits (Patching, 1996).

One benefit that is particularly important for this economic assessment is the role of parks as a powerful magnet for wildlife and tourists. The world-famous Etosha National Park and the dunes at Sossusvlei in Namib-Naukluft Park attract tourists from all over the world to Namibia, while the network of protected areas provides focal points for both tourists and wildlife across the country. Without the protected areas, the economic benefits generated from wildlife on the non-protected land, and in the tourism industry more broadly, would be lost.

Regional Magnet and Motor

The function of parks as regional magnet and motor is already evident in the mushrooming of private game reserves on the southern border of Etosha and eastern border of the Namib-Naukluft Park. A further indication of these ben-efits comes from the research discussed previously on the economic value of wildlife uses on communal land. Areas that are adjacent to protected lands are much more valuable than those further away. In the northwest study areas, the highest current and potential economic benefits per square kilometer are in the areas adjacent to the Skeleton Coast Park and Etosha. Economic benefits in these areas could increase by around 300 to 400%, depending on how much the resource base is expanded, compared with increases of 80 to 160% in areas further away.* These results show that parks are adding value to neighboring areas. To exploit this potential, multiple-use buffer zones should be developed in which wildlife use dominates other uses.

Tourism facilities on the edges of protected areas are not only benefiting from their proximity to a tourism destination, but in many cases the maintenance of wildlife habitat, and hence viable wildlife populations inside protected areas, also makes possible the dispersion of wildlife beyond the park into communal or commercial land. An indication of the value of this free-ranging asset can be gleaned from the financial analysis of wildlife-viewing game lodges described previously. The financial return on a game lodge is low because of the massive N$3.2 million investment it entails, of which 38% is the cost of stocking up with wildlife. Therefore, those lodges that enjoy some natural dispersion of wildlife

---

* Estimated current economic benefits average around N$41 per square kilometer in zones adjacent to protected areas, compared to N$22 for non-adjacent areas. The potential values are N$125 compared to N$39 per square kilometer, and with improved resource stocks N$170 compared to N$57, in adjacent and non-adjacent areas, respectively.

onto their land from adjacent protected areas can achieve higher profitability. From the national economic point of view, this natural dispersion saves economic costs which are necessary if game has to be moved from one part of the country to another.

National Magnet for Tourism

By attracting tourists to Namibia instead of other holiday destinations, the national parks are providing a foundation for Namibia's tourism industry. The vast majority of overseas holiday tourists visit Etosha and Sossusvlei. In 1993 and 1994, over 40,000 overseas holiday-makers visited Namibia, of which 25,000 to 30,000 went to Etosha and 20,000 went to Sossusvlei.* These tourists are clients for the tourism facilities in the communal and commercial areas discussed above and also for tour operators, car or plane hire companies, restaurants, taxis, airlines, souvenir sellers, and so on. Indeed, it is estimated that tourists spend just as much on these other items as they do on accommodations and viewing wildlife (Hoff and Overgaard, 1993). It is therefore necessary to consider the overall value of tourism in the national economy when assessing the contribution of national parks.

The total expenditure by wildlife-focused tourists was estimated at over N$350 million in 1992, which indicates that the contribution to net national income from wildlife-based tourism was almost N$200 million** (roughly N$250 million per year in 1994 prices). This can be seen as the net economic benefit of the industry for which wildlife and scenery in national parks and reserves is the core resource.***

---

* Assumes that around 80% of overseas (non-African) tourists visit for leisure purposes and that the average number of nights per person spent in Etosha is 2.5 (Hoff and Overgaard, 1993, and unpublished data). The percentage of African tourists visiting Etosha is smaller, probably because they are on repeat visits or visiting friends.

** Total expenditure by international and domestic tourists was N$509 million in 1992. Estimates assume that 60% of tourists are wildlife focused, 29% are business tourists, and 10% are visiting family and friends. Estimates also assume that the wildlife-focused tourists account for 70% of tourism expenditure because they stay longer (Hoff and Overgaard, 1993) and that net economic contribution is equivalent to 55% of turnover. Estimates are inflated to 1994 prices using the Windhoek Consumer Price Index (Ministry of Finance, 1994).

*** However, parks and reserves also have a negative effect on the tourism industry in that the subsidized prices of government accommodation affect the competitiveness of private tourism establishments outside parks. The resulting reduction in demand for, and prices of, private accommodation has not been quantified, although it may well diminish in the foreseeable future as the commercialization of government resorts will require cost recovery and, doubtless, price increases.

Overall Economic Value of Wildlife and Tourism

Table 8.2 fits the pieces together to give a rough picture of the economic value of direct uses of wildlife and tourism in Namibia. It must be remembered that other benefits of wildlife, indirect and non-use values, are not quantified. Although the figures are approximate, it is clear that the benefits are currently concentrated in commercial rather than communal land and that the potential for non-consumptive tourism benefits to outweigh the consumptive benefits, particularly on communal land, was not yet realized in 1994. It is also noteworthy that the economic value of supporting services for the tourists who come to enjoy wildlife and wilderness is even greater than that of the direct wildlife-using enterprises. Given a potential doubling of tourism arrivals by 2000, according to the Tourism Development Plan (Government of the Republic of Namibia, 1995; Hoff and Overgaard, 1993), the devolution of rights over wildlife to conservancies in communal areas, and the ongoing expansion in wildlife and tourism on commercial land, a doubling of these estimated economic ben-

Table 8.2   Overview of Estimated Net Economic Contributions of Wildlife Utilization Activities in Parts[a] of Namibia (in Millions of 1994 Namibian Dollars)

	Non-consumptive tourism	*Consumptive uses*	*Total*
Commercial land[b]	15–20	32–37	52
Northwest and northeast communal land[c]	5[d]	3–4[e]	8–9
Total	20–25	35–41	**60–61**
Additional services for wildlife-viewing tourists			**190[f]**

[a] Economic benefits of national parks are not estimated, although part of their value is captured in the last row.

[b] Source: Barnes and de Jager (1995) and Department of Environmental Affairs unpublished data.

[c] Accounting for 43% of all communal land, but most of the remainder has considerably lower wildlife potential. Source: Barnes (1995), supplemented by further estimates for trophy hunting on communal land (Barnes, 1996).

[d] Includes craft production and sales, as tourists are the primary market.

[e] Hunting and angling by tourists, plus local harvesting of wildlife and freshwater fish for subsistence or local sale.

[f] The estimated economic contribution of wildlife-viewing tourists in 1992, inflated to 1994 prices (N$250 million), less N$60 million generated directly from enterprises on communal and commercial land.

efits is easily foreseeable. If the natural resource base is enhanced and tourism developed sustainably, greater increases are possible.

Benefits to Local Residents on Communal Land

The development process has as an important objective: improved livelihoods and opportunities for the historically marginalized poor who make up the majority in communal areas. Wildlife utilization boosts the economy, but who benefits? How significant is it to the residents of communal areas, who suffer the costs of wildlife damage, who live in the areas visited by tourists, and who are expected to conserve wildlife? It is of crucial importance to find strategies through which wildlife use for economic gain benefits rural communities, for the sake of both development and conservation. Until recently, residents in communal areas had almost no rights to utilize wildlife and few opportunities to participate in the historically white tourism sector; thus, financial benefits for local residents have been confined mainly to wages in private tourism enterprises. But new developments are changing this.

- Communities are gaining rights to use wildlife and develop tourism through conservancies.
- Communities and local residents are initiating their own tourism enterprises and entering partnerships with the private sector.
- Prime areas for the most profitable up-market ecotourism developments fall within communal land.

The most fundamental requirement for ensuring that local communities can derive benefits from wildlife is appropriate property rights. Individual rights of tenure over wildlife are not feasible in communal areas because of the social structure, relatively high human densities, and large areas needed by most animal species. However, the new policy and pending legislation will permit the development of conservancies and thus common property control and management of the wildlife resources. This will include the right to prevent open access to wildlife resources, to manage them for maximum gain, to charge for access to wildlife, and to accrue marketable assets in the form of wildlife stocks.

Research and analysis by Ashley and Garland (1994) and Ashley (1995), which build on the work in the four study areas of Barnes (1995), show that there is potential for local net incomes* earned from wildlife to triple in the northwest

---

* Net incomes here may be defined as take-home wages, royalty/profit-sharing payments to communities, or net profits from community or individual resource use activities.

and northeast communal areas, even without any increase in the resource base. However, the significance of this for rural development depends on many factors, including the type and distribution of benefits and their scale compared to population density and alternative incomes, as the following sections show. In turn, implications for maximizing the positive impact of wildlife use can be identified.

Different Types and Distribution of Benefits from Wildlife

Wages

Different enterprises will provide very different levels of financial and other benefits to residents of communal areas. As Table 8.3 shows, local wages from an up-market lodge can be up to N$80,000 per year and are the most substantial financial injection into the local economy. Wages outweigh what a community could earn from its own enterprise or from a voluntary bed-night levy and might only be matched by a concession fee earned by a conservancy from a joint venture.

Locally Controlled and Distributed Income

In East Caprivi, several up-market lodges inject a few hundred thousand dollars of wages into the local economy, and one lodge, Lianshulu, pays a voluntary bed-night levy to its neighbors of around N$15,000 per year. But it is the bed-night levy that has focused attention on the benefits of wildlife conservation and that has required conflicting communities to establish procedures for sharing the money; for hundreds of households, it also provides their first-ever cash benefit from wildlife: N$35. In terms of the development impact of wildlife benefits, it is not just the amount of cash that matters but how it is distributed and who decides.

The bulk of local income from wildlife will never be shared equally among rural households because it comes in the form of staff wages. Jobs in lodges and camps are bound to go to those most skilled or nearby, and their allocation is decided by an outsider. Earnings of local artisans (craft-makers or guides) will also depend on the distribution of skills. However, collective income can be earned by a community from its own enterprise (e.g., campsites, craft centers), bed-night levies donated by private operators, meat from a hunt, or concession fees paid to conservancies (a few thousand dollars per year in the first three of these cases or tens of thousands of dollars in the last, as shown in Table 8.3). This collective income is qualitatively different from wage income, because it can be locally controlled and more broadly distributed.

Table 8.3 Benefits and Costs (in Namibian dollars, K = 1000) to Local Residents of Selected Wildlife-Based Enterprises on Communal Land

Benefits/ costs	Enterprise						
	Private lodge, up-market tourism	Private lodge, voluntary revenue share	Joint-venture lodge (private + community)	Community tourism enterprise	Hunting camp in govt. concession	Hunting camp in conservancy	Local wildlife cull by residents
Financial, p.a.							
Local wages	N$50K–80K	N$50K–80K	N$50K–80K	N$500–1000 per craft household	N$44K	N$44K+	None
Collective income	None	N$15K–20K[a]	N$40K–80K[b]	N$2K–20K+ per community	Meat worth N$6K	N$100K±;[c] some meat	Meat worth N$50K[d]
Social							
Skill and institutional development	None	Some, in revenue distribution	Some: negotiation and distribution; possibly mgmt.	Some mgmt. and distribution	None	Some negotiation and distribution; possibly mgmt.	Some culling and distribution possibly mgmt.
Local rights control ownership	None	Some control of revenue but no rights or ownership	Some rights, control of revenue; possibly some ownership	Some, perhaps if not privatized by an individual	None	Some rights; possibly some control, ownership	None, perhaps unless inside conservancy[e]
Costs to community (excl. wildlife damage)	Loss of land and resources	Loss of land and resources	Difficult; time and effort, risk of failure	Difficult; time and effort, risk of failure	Loss of land rights and resources	Time and effort for negotiation	Time and effort in hunting

a For example, a N$5 bed-night levy for a lodge charging around N$200 per night or N$10 for a more exclusive but smaller lodge charging N$400 per night. Based on generalized enterprise models, these are estimated to be viable for a lodge operator, particularly if the levy boosts tourist appeal or wins reciprocal local benefits (e.g., Lianshulu Lodge collected in 1994 and part of 1993).

b For example, a N$25 bed-night levy from an up-market camp or 5 to 15% share of turnover (15 to 50% of profit). This is viable for the operator if communities can offer some security on land/wildlife/tourism assets, such as in a conservancy, and/or the lodge attracts "ethical tourists."

c Very variable depending on conservancy size. Assumes average conservancy is half the size of current hunting concessions and all the concession fee is paid to the conservancy rather than the government.

d For example, in Sesfontein area in 1993, three local hunts produced 42,000 kg of meat valued at $3.50 per kilogram, giving an average value per area of N$50K. Profits from sale of skins and costs of ammunition are not shown. The profits potentially outweigh costs, enabling cash income to be generated in addition.

e To date, local hunts are controlled and supervised by the Ministry of Environment and Tourism.

Apart from any moral preference for equity, there are important practical reasons to value the local control and broader distribution of benefits of wildlife on communal land. From a development perspective, impacts on living standards and poverty alleviation are likely to be greater if benefits reach the poorest households. In addition, the development of skills and institutions required to distribute revenue can boost other local developments.* From a conservation perspective, it is important that benefits reach all rural residents in wildlife areas of communal land, because if the majority of people on communal land remain committed exclusively to livestock, or even a minority to poaching, then collective wildlife management breaks down. Apart from the financial benefits, participation in the *management* of resources has proven immensely important in the success of community-based conservation projects in Namibia, and community control of revenue is one important part of participation. In the long term, this community commitment to conservation is essential if all the other national and local economic benefits discussed in this chapter are to be achieved in communal areas.

Social Benefits

Social benefits, such as the development of skills and institutions, may be gained from tourism enterprises in ways other than through control of money. In particular, entrepreneurial skills are more likely to develop in community enterprises and joint ventures, and a sense of empowerment is more likely to develop from enterprises controlled by communities. However, social costs also need to be taken into account. Joint ventures in particular require enormous time and effort (transaction costs).

This analysis has implications for the type of wildlife use promoted in communal areas, as it suggests that the "value" of community-controlled income from bed-night levies, hunting or tourism concessions, or community enterprises is higher than reflected in dollar terms. In economic terms, it implies a weighting for these locally controlled earnings. In addition, other development benefits and costs need to be taken into account in any cost–benefit analysis. It is also important to seek a combination of enterprises and increase the upstream and downstream linkages of any development. Many up-market wildlife-viewing lodges are linked more closely to Windhoek or Johannesburg than to the local

---

* There are also costs to local distribution of benefits in terms of time and effort needed to arrange distribution. In Caprivi, preparations for the Lianshulu bed-night levy distribution occurred over a year. Delays are common and there is a risk of mismanagement.

economy, but they could be the focus for a network of secondary enterprises ranging from firewood and laundry to cultural shows and home visits.

The Scale of Financial Benefits at the Regional and Household Levels

Regional Current and Potential Benefits

Looking at the bigger picture, how much can wildlife utilization contribute to local incomes overall in Namibia's communal areas? Barnes's (1995) study shows that residents of the northwest and northeast communal areas* are currently earning around N$2.1 million from wildlife enterprises. Wages of local staff employed in wildlife-viewing lodges account for half of this, while the production and sales of crafts account for a quarter. With expansion up to sustainable limits and no increase in the natural resource base, local income could triple to N$6.8 million per year.

Comparisons between regions and between different types of enterprises follow a pattern similar to the estimates of net economic contribution described above (see Figure 8.3). Caprivi, relatively well endowed with natural resources and tourism infrastructure, enjoys the highest absolute level of current and potential income while remote former Bushmanland has the lowest level of income, but highest potential rate of increase. Potential is also highest in areas adjacent to protected areas. Non-consumptive tourism again dominates the picture in the arid northwest, where carrying capacity is low but scenic quality high. This is in marked contrast to some other community-based conservation programs, notably CAMPFIRE in Zimbabwe, where hunting provides the bulk of community benefits.

With a potential tripling of local staff wages from tourist camps and lodges to around N$3.5 million, wages still account for over half of potential local income. But the critical question is whether communities' revenue shares, royalties, and concession fees from tourism and hunting operators develop to a similar scale. Voluntary revenue sharing by lodges on a broad scale could generate up to N$1 million in total for local communities, but if conservancies are established with concessionary rights to virtually all prime sites outside protected areas, lease fees could total around N$3 million once normal turnover levels are achieved (Department of Environmental Affairs, unpublished data).

---

* All aggregate income estimates in this section are derived from Barnes (1995) and apply only to the four study areas: Caprivi region, former Bushmanland, former Damaraland, and Opuwo. More detailed analysis of returns to different activities, zones, per capita, and per hectare are from Ashley (1995) derived from Barnes (1995).

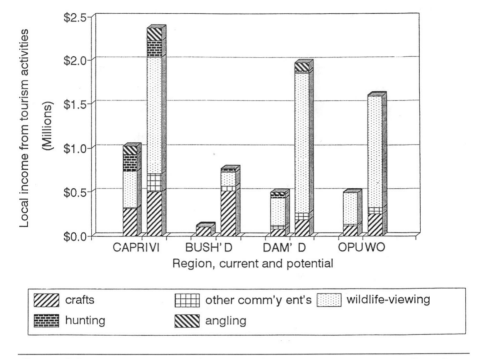

Figure 8.3   Current and potential local income (in 1994 Namibian dollars) from wildlife use in four study areas in communal land, by type of activity.

Earnings per Household in High- and Low-Potential Zones

Caprivians will not take heart that their region enjoys highest total earnings from wildlife and tourism if the amount *per household* is insignificant. Indeed, potential wildlife income per resident is higher in "former Damaraland," where the population density is considerably lower.* In zones with *medium wildlife potential*, average wildlife income per household could increase from N$100–200 per

---

* Within the region, the areas with highest potential tend to be least populated and vice versa, so there are vast differences in income per person among the zones, ranging from only N$2 per resident in the more populated and less scenic eastern end of the Ugab to nearly N$5000 per resident in the upper Uniab catchment, with low population and high tourism capacity. Excluding these extremes, wildlife and tourism in most zones could generate N$100 to N$230 per person per year, with generally higher potential in areas adjacent to the Skeleton Coast Park and Etosha. The average for the entire region (i.e., if all the estimated local wildlife income were spread equally across the population, which is highly unlikely) is N$15 per resident per year now and N$60 in the potential scenario (with no increase in the resource base). Figures per household in the text assume an average household size of ten.

year to N$500–N$1000 in Caprivi* and to N$1000–N$2000 per year in former Damaraland. Increases would be somewhat greater if the resource base improves. Of course, in practice some households will receive well above average and others below. It is estimated that average household income of subsistence farming households is around N$7000 per year and around N$2000 per year in the poorest 20% of households (Central Statistics Office, 1995a). On this basis, wildlife enterprises could make a substantial contribution to household incomes but not revolutionize them.

However, in zones with *high potential*, which are generally adjacent to protected areas, the order of magnitude is nearer N$10,000 per household per year or more, making wildlife and tourism very important development options. This illustrates the importance of prioritizing developments in the high-potential areas.

Benefits vs. Costs of Wildlife

Caprivi's elephants relish a midnight feast of green "mealies" (maize, corn-on-the-cob) still growing on the cob, about to be harvested. Kunene's elephants will follow the smell of water and dig up pipes and pumps if they find the ground dry. For lions and hyena, goats and calves not herded in at night can be an easier catch than antelope. The residents of communal areas suffer these costs, and not surprisingly, many see wildlife as nothing but trouble. What benefits must wildlife generate to be perceived instead as a route to development?

Four years of research in one of the areas worst affected by wildlife damage, the east bank of the Kwando River in east Caprivi, found that between 1991 and 1994 the 13 most affected villages lost an average of N$1000 worth of crops per village per year due to elephant damage (O'Connell, 1995). Losses of livestock to predators cost about another N$2000 per village, except for the four villages bordering Mamili National Park, where lion attacks are more common and losses higher. This means that for most communities, a very small enterprise (a one-fifth share of an annual bed levy or one employee in a lodge) would provide income on a comparable scale with direct losses.

In fact, in the wildlife damage study area, it is estimated that workers in tourism lodges and craft makers already earn around N$300,000 per year in total—four times the estimated agricultural losses from wildlife of around N$70,000 per year since 1991 (Ashley and O'Connell, 1996). If wildlife uses

---

* In Caprivi region, the range goes from a low potential income per capita of around N$30 in several of the zones up to N$100 to 300 in prime riverine areas. Across the whole Caprivi region, current wildlife tourism income averages around N$15 per capita and potential income around N$35 per capita per year.

increase to potential, total local income could be eight or ten times the cost of wildlife. However, residents of the area certainly do not *perceive* that benefits of wildlife are already four times greater than damage costs. This is doubtless because tourism income is likely to be *distributed and perceived quite differently* from the damage costs of wildlife. However much is earned by neighbors with tourism jobs, it is still perfectly possible that losses for individual farming families may exceed any benefit, unless collective income is very large and equally shared. Furthermore, wages and other tourism income are less public and the link with wildlife not demonstrated and discussed. The link between lost crops and elephants is all too evident and vocally discussed the morning after each event.

These observations suggest that the benefits of wildlife are more likely to exceed the costs at the household level, in perception and in practice, under three conditions: (1) if benefits of wildlife are broadly distributed between households and at least a share can be allocated by communities themselves in accordance with their perception of fairness; (2) if links between tourism income and wildlife are emphasized; and (3) if, in the aggregate, benefits exceed costs sufficiently that households receiving below-average benefit get enough to match their agricultural losses (i.e., benefits need to be either massive or very evenly distributed and well publicized). Otherwise, the majority are unlikely to invest in wildlife as a rural development strategy.

Wildlife as a Complement to Other Land Uses

Livestock, crops, and a range of natural resources, such as grass, wood, and fruits, provide the essentials of life for most rural households in communal areas, plus the means to earn some cash from local sale. In deciding whether to switch time, effort, and, most importantly, land to wildlife and tourism, households will compare the returns on these various activities and decide on an appropriate combination. The question is not whether wildlife can provide enough to become the *only* option but whether it could become a major addition to livelihoods, and hence a viable constituent land use.

Comparisons with the harvesting of other wild resources, such as thatching grass, palms, reeds, fish, fuelwood, and timber, are difficult because subsistence use is hard to value and quantify, but rough comparisons by Barnes (1995) in the four study areas show that local income from harvesting of non-agricultural resources currently exceeds income from wildlife in the northeast, whereas the reverse is true in the more arid northwest. However, because wildlife income has greater potential for expansion, in the future this income could be dominant in all four areas. It is most likely that the greatest opportunities for households to *increase* income come from wildlife and tourism.

A comparison with livestock agriculture is more important than a comparison with wild plant resources because, to some extent, wildlife tourism competes with it. Wildlife and livestock can and do share habitats, but some limited zoning of land is necessary to provide high-quality core areas for tourists and wildlife and reduce the damage of livestock by wildlife. The value of setting aside land will depend on the returns per hectare of competing activities.

A joint venture up-market lodge in a communal area could generate local income (wages plus revenue share) of N$150,000 per year. If this income is attributed just to the 4-ha lodge site, the return to land is massive. If, more reasonably, it is attributed to land use of the whole concession area, say 14,000 ha, the return per hectare is over N$10/ha, which is still good. However, the viability of the lodge may depend on maintaining wildlife over a much larger area of thousands of square kilometers. If the tourism income is averaged across the entire region, the return seems much less impressive. For example, potential tourism income averages out at N$1.2/ha across Caprivi and N$0.33/ha in former Damaraland.

This implies that at specific sites, particularly prime tourism sites, it could be very profitable for a community to substitute wildlife for agriculture. On a broader scale, however, it will be a complement to agricultural income, not a substitute way of life. This complementarity depends on maintaining wildlife stocks across the larger area (i.e., maintaining multiple-use zones where livestock and wildlife coexist). Therefore, the priorities are to identify:

1. Which sites would be better used for tourism than agriculture
2. The extent to which wildlife and agriculture can complement each other across large farming or residential areas and ways to reduce the trade-offs
3. Ways to ensure that wildlife *is* protected in the larger mixed-use areas through sufficient incentives and opportunities

At the same time, the core conservation areas need to be maintained, as they act as the magnet and motor for tourism development opportunities in the communal areas.

In summary, the diversification of income and risk is a way of life for the poor, and wildlife adds another element to this survival strategy. Tourism enterprises cannot provide the basics of maize, meat, and milk, but can provide a little *cash income,* which is so essential for school expenses, clothing, sugar, and other marketed goods. Furthermore, because non-consumptive tourism is relatively independent of drought cycles (though subject to other fluctuations), it can *dilute risk* and act as a drought buffer. A final and important indicator of the significance of wildlife for rural development is the potential for *increases* in local wildlife incomes, which are probably greater than potential increases from agriculture.

Overall Contribution of Wildlife Enterprises to Development in Communal Areas

While the economic value of wildlife and tourism on communal land is relatively small in the national perspective, it is clear that it can be highly significant for local development and improved living standards in some of the most marginalized areas of the country in the northwest and northeast communal areas. It can boost and diversify local incomes substantially in many areas, providing a complement to agriculture on a large scale and a highly profitable alternative at prime sites. As residents become more involved in tourism, social benefits such as increased skills and institutional development are also likely. These economic and social benefits can, in turn, increase people's commitment and capacity for conservation, and thus enable further growth of the industry and its benefits.

The impact of development impact will depend on the type of wildlife enterprises as well as their scale. It will also depend in particular on the degree to which communities are involved and how they distribute benefits. Trade-offs are likely between maximizing incomes and rate of development, through private sector operations, and increasing community participation and control, and through local enterprises and initiatives. Conservancies, within which communities can lease out concessions to private operators, are an ideal way of combining private sector money and expertise with local control, but such joint ventures will not happen overnight. For wildlife to be broadly perceived as a development option by local residents of communal areas, equitably distributed collective income will be needed, in addition to jobs for a minority, emphasis on links between tourism income and wildlife, and minimization of conflicts with livestock.

Although tourism is developing rapidly in communal areas, the framework is not yet in place to maximize development impacts of wildlife use. There is a lack of tourism planning to ensure that prime areas are neither under- nor over-utilized, no obligation on existing tourism or hunting operators to involve communities, conservancy rights are only just being legislated, and communities lack a range of skills and even basic information for operating wildlife enterprises. However, given the potential benefits that can be realized, action is being taken by the Namibian government, non-governmental organizations, and donors to address these obstacles.

Conclusions

Namibia has good potential for expansion of sustainable wildlife use, which can contribute positively to national economic growth and more than double its economic value over the next 10 to 20 years. In some communal areas, local incomes from wildlife could increase severalfold within that time. However, within the context of Namibia's dual economy, there remain stark differences in capacity to achieve this potential. In the commercial farming sector, most necessary conditions are in place to ensure growth of wildlife stocks and use, in particular appropriate property rights for commercial farmers. They are already diversifying from livestock and gradually moving to suitably profitable uses of wildlife. However, in the communal lands, where most of the people live and where most are poor, much still needs to be done to ensure growth in wildlife stocks and use.

Namibia's most intrinsically valuable wildlife resources are found in or adjacent to the communal lands. Without the appropriate conditions, these will be lost—the space they occupy will be converted to agricultural uses and their stocks depleted. What is required is high tangible use values for wildlife, realizable by communal land residents, within appropriate property rights. The preceding discussion highlights certain strategic principles to achieve this. In order for communal area residents to manage, benefit from, and invest in wildlife, Namibia should:

- Develop community rights
- Support conservancies and local enterprises
- Seek to develop wildlife as a complement to agriculture and minimize trade-offs
- Make sure wildlife use rights include non-consumptive uses, given the high potential for tourism as well as consumptive uses

These goals will require concerted efforts by communities, government, nongovernmental organizations, and the private sector. Getting enterprises going is just the beginning; how they are implemented and controlled, and how benefits are shared, will really determine the contribution to local development and conservation. However, the evidence so far shows that the boost to the national economy and to development in marginalized communal areas will make it a worthwhile investment.

Acknowledgments

This work was undertaken within the Directorate of Environmental Affairs, Ministry of Environment and Tourism, Namibia. The work of the resource economists was supported by the LIFE (Living in a Finite Environment) Programme of the World Wildlife Fund (WWF–US), sponsored by USAID, and by the Overseas Development Institute (ODI), London. We thank all those who provided support and assistance, particularly C.J. Brown, B.T.B. Jones, J.L.V. de Jager, L.C. Weaver, B. Wyckoff-Baird, and T. Healy for support and assistance.

References

Ashley, C. (1995) Tourism, Communities, and the Potential Impact on Local Incomes and Conservation. Research Discussion Paper No. 10. Directorate of Environmental Affairs, Ministry of Environment and Tourism, Windhoek, Namibia.

Ashley, C. and Garland, E. (1994) Promoting Community-Based Tourism Development: Why, What and How? Research Discussion Paper No. 4. Directorate of Environmental Affairs, Ministry of Environment and Tourism, Windhoek, Namibia.

Ashley, C. and O'Connell, C. (1996) Can the financial benefits of wildlife outweigh the costs for Caprivi households? In: *Namibia Environment,* Volume 1. P. Tarr, Ed. Ministry of Environment and Tourism, Windhoek, Namibia, p. 194.

Barnes, J.I. (1995) The Value of Non-Agricultural Land Use in Some Namibian Communal Areas: A Data Base for Planning. Research Discussion Paper No. 6. Directorate of Environmental Affairs, Ministry of Environment and Tourism, Windhoek, Namibia.

Barnes, J.I. (1996) Trophy hunting in Namibia. In: *Namibia Environment, Volume 1.* P. Tarr, Ed. Ministry of Environment and Tourism, Windhoek, Namibia, pp. 100–103.

Barnes, J.I. and de Jager, J.H.V. (1995) Economic and Financial Incentives for Wildlife Use on Private Land in Namibia and the Implications for Policy. Research Discussion Paper No. 8. Directorate of Environmental Affairs, Ministry of Environment and Tourism, Windhoek, Namibia.

Brown, C.J., Ed. (1994) *Namibia's Green Plan.* Ministry of Environment and Tourism, Windhoek, Namibia.

Central Statistical Office (1995a) Living Conditions in Namibia, Main Report, Part 1, of the 1993/4 Namibia Household Income and Expenditure Survey (Draft). National Planning Commission, Windhoek, Namibia.

Central Statistical Office (1995b) The Distribution of Economic Resources in the Population of Namibia. National Planning Commission, Windhoek, Namibia.

Central Statistical Office (1995c) National Accounts 1987–1994. National Planning Commission, Windhoek, Namibia.

Gittinger, J.P. (1982) *Economic Analysis of Agricultural Projects.* Johns Hopkins University Press, Baltimore, MD, 505 pp.

Government of the Republic of Namibia (1995) First National Development Plan: 1995/1996–1999/2000 (2 volumes). National Planning Commission, Windhoek, Namibia.

Hoff and Overgaard Planning Consultants Ltd. (1993) Namibia Tourism Development Study, Main Volume. Ministry of Wildlife Conservation and Tourism and Commission of the European Communities, Windhoek, Namibia.

Joubert, E. (1974) The development of wildlife utilization in South West Africa. *Journal Southern African Wildlife Management Association,* 4: 35–42.

Joubert, E. and Mostert, P.M.K. (1975) Distribution patterns and status of some mammals in South West Africa. *Madoqua (Namibia),* 9: 4–44.

Ministry of Finance (1994) Economic Review, 1994. Ministry of Finance, Windhoek, Namibia.

National Planning Commission (1994) Namibia: Population and Development Planning. National Planning Commission, Windhoek, Namibia.

O'Connell, C. (1995) Final Technical Report; East/West Caprivi Natural Resource Monitoring Project: Elephant/Human Conflicts. Living in a Finite Environment (LIFE) Programme, Windhoek, Namibia.

Patching, N. (1996) Costs and Revenues from Game Parks. Namibia's Biodiversity Country Study. Ministry of Environment and Tourism, Windhoek, Namibia, forthcoming.

Tapscott, C. (1992) Namibia: A Technical Background Paper. Inter-Agency WCARRD Policy Review Mission to Namibia, Social Sciences Division, Multidisciplinary Research Centre, University of Namibia, Windhoek, Namibia.

World Bank (1992) *Namibia: Poverty Alleviation with Sustainable Growth: A World Bank Country Study.* The World Bank, Washington, D.C.

Yaron, G., Healy, T., and Tapscott, C. (1994) The economics of wildlife in Namibia. In: The Economics of Wildlife, White Cover Report. J. Bojo, Ed. Environmentally Sustainable Development Division, Technical Department, Africa Region, The World Bank, Washington, D.C., pp. 49–86.

Resources Used in Attaining Land Tenure Security: The Case of Peruvian Amazonia

9

Roxana M. Barrantes
Departamento de Economía, Pontificia Universidad Católica del Perú, Apartado 1761, Lima, Perú

Abstract

It is commonly argued that before individuals can invest in making a living from the land, their security of tenure must first be established. This chapter shows that the reverse is true in Amazonia: investments of time and money are made in order to achieve land tenure security. This situation is a function of the regulations governing the granting of titles. In Peru, these regulations require evidence of significant investments by individuals and the use of the land's resources to prove occupation. These regulations are, however, only weakly enforced by the state. The case studies presented in this chapter document conflicts over land access and the resources—natural and otherwise—consumed in the process of gaining exclusive access to it. Policy recommendations for the sustainable use of Amazonia focus on changing the distortionary effects of current Peruvian law and on international cooperation.

1-57444-077-2/97/$0.00+$.50
© 1997 by CRC Press LLC

193

Introduction

The recent surge of market-oriented policies and privatization around the world has brought about an emphasis on property rights in less developed countries. Land tenure, privatization, and accompanying titling projects are being put into effect particularly in Africa and Latin America. Substantial amounts of borrowed funds are allocated to such efforts with little empirical knowledge as to their possible real impact, and consequently, the expected benefit stream from those projects may never be realized.

Frontier areas are by definition where the enforcement powers of the state are weak or non-existent. As a result, the rights to appropriate the benefits from resource use—property rights—are still not fully defined and are usually backed by private enforcement. Individuals on the frontier consume a variety of resources in the process of defining and enforcing property rights, including rent dissipation in the form of violence to enforce claims, time spent in some form of governmental recognition, bribery to initiate or accelerate bureaucratic procedures, and ecological degradation through rapid exploitation of resources. The possession of a title to land under these conditions may not necessarily have any importance.

Peruvian Amazonia is one such frontier area. One policy option to stop deforestation and prevent resource degradation is to privatize the land, because owners are supposed to have the greatest incentive to realize value from the land. In frontier areas, however, it may well be that land-specific investments are a *prerequisite* for tenure security, which is attained by clearing the land and using it for agricultural purposes, after which formal tenure may be demanded (but not before). The objective of this chapter is to examine empirically the problem of attaining tenure security in a frontier area—Peruvian Amazonia—to shed light on the possible effectiveness of titling projects to address issues of environmental degradation. The focus is on institutional restrictions because they shape the way resources are accessed and legally sanctioned.

The following section contains a review of the relevant literature, followed by an analysis of the institutional constraints to securing land tenure in Peruvian Amazonia, including descriptions of individual cases. Policy recommendations are offered in the final section.

Review of the Literature

There are a number of studies in the literature on the formation of private property rights for resources and, related to those, on frontier conditions in the Western United States. Studies about Amazonia concentrate on Brazil, and the particular issue they deal with is land conflicts.

On the Formation of Private Property Rights

The Western United States

Using the 19th century North American West as the empirical test of their theories, Anderson and Hill (1975, 1983, 1990), Umbeck (1977, 1981), and Stroup (1988), among others, developed a series of testable hypotheses about the creation of property rights under frontier conditions.

In their studies, Anderson and Hill emphasize the process by which property rights come into being. The importance of the dynamics of this process lies in the resources that are consumed and, consequently, in the possible dissipation of rents that the privatization of the commons may generate. In their 1975 paper, Anderson and Hill emphasize changes in the activities of defining and enforcing property rights, as well as the timing of those changes. Product demand, factor endowments, and technology are postulated to be at the basis of those changes. Because property rights decisions are made at the margin, Anderson and Hill's analysis goes in the direction of identifying the nature of the benefit and cost functions for the definition and enforcement of property rights. These are inputs into the granting of property rights instead of the finer concept of "degree" of private property (i.e., how effectively one maintains exclusivity). Although it is recognized that the latter is the important concept for resource allocation, it is also difficult to measure.

Anderson and Hill (1983) discuss the possible rent dissipation created by privatization when it is imposed exogenously by non-claimants rather than devised and enforced endogenously by residual claimants. Residual claimants will tend to economize on definition and enforcement activities because they are the sole bearers of costs. Hence, rent dissipation in the course of privatization will tend to be lower. For the American West, the implication is that the definition and enforcement of property rights were different under the Homestead Acts than when rights were devised according to the interests of the contracting parties. They illustrate the theory by comparing claims associations that provided definition and enforcement vis-à-vis those accessing lands under the provisions of the Homestead Acts, which explicitly required expenditures in labor and capital in order to establish ownership.

The following papers addressed the issue of the timing of establishing property rights and bringing resources into production:

1. Southey (1978, p. 554) developed a simple model of the timing of bringing land into agricultural production when free access to land prevails. "The date of commencing operation is then chosen so that the expected net annual rent at that time just covers the interest charge on the set-up costs, plus any appreciation or depreciation of the set-up costs over time." In his model, the set-up costs, however, do not include the costs of

property rights definition and enforcement. The model predicts that when open access to resources prevails, both homesteading and farming will begin before the optimal time.

2. Stroup (1988, p. 69, 75) discusses these issues using the theory of rent seeking. He predicts that competition for land generates premature settlement and a loss of settlers' wealth equivalent to all the potential value embodied in the unsettled territory beyond the frontier. However, if settlers are heterogeneous, competition will not raise the costs so as to eliminate all possible rents. Stroup postulates that if claimant heterogeneity rests on different expectations about future returns to ownership of each parcel of land, a sort of "winner's curse" takes effect, where the highest bid will overvalue the resource and dissipate more rents than when claimants are homogeneous. Under this winner's curse, competition will "encourage settlement well before the settlers really wanted to, but also systematic errors (a self-selection bias) on the part of those settling first would compound the problem."

3. Anderson and Hill (1990) extend Stroup's model to incorporate the costs of definition and enforcement of property rights. In this way, they analyze the effect of homesteading, squatting, and speculating on the timing of property rights definition and bringing resources into production. Under speculation—defined by the authors as sale by the government of land at minimum price—land can be purchased before the optimal time to put it into farming, but actual farming will be delayed until that optimal time. The establishment of both property rights and farming will be fastest under homesteading. With squatting, the definition of property rights takes place before or when farming commences, but this process will be faster than with speculation.

4. Last but not least, Umbeck (1981) presents a model of the initial formation and distribution of property rights, singling out violence as a major constraint on this formation process. Violence, or the threat of violence, is important because ownership rights are based on the abilities of individuals to forcefully maintain exclusivity. Violence is understood as the labor time allocated to excluding other individuals from a piece of land. Under the assumption that labor is homogeneous, when land is homogeneous, the model predicts that land will be divided equally. If land is heterogeneous, the most productive land will be held in smaller plots than less productive land. The theory is tested for the California gold rush.

In summary, by focusing on the North American West in the 19th century, these studies compare different forms and timing of access to land with or without government regulation and enforcement. The emphasis is on the costs of defining and enforcing private property rights (i.e., gaining exclusivity to resource use) and the potential rent dissipation that a process of privatization creates. The alternative resource uses open to claimants as a consequence of the way land was accessed are not specifically discussed, although it is noted that homesteaders may have suffered losses by being required to make land-specific investments to claim exclusivity.

Two issues stand out as relevant for this chapter. First, it is important to identify the conditions under which the activities of definition and enforcement of property rights are demanded on the frontier. Second, the regulations imposed by the government on how exclusive resource access is obtained, and the amount of resources that it consumes, are very important as well. In the absence of regulations, claimants have more choices, and the alternative to directly contract and assume private enforcement is unimpeded. With regulations—in the direction of investment in land to gain access—additional constraints are imposed on settlers, which may involve higher rent dissipation than without government regulations.

Brazil

Due to both the aggressive policies that the Brazilian government carried out to occupy the region and the vastness of Amazonia under its control, Brazilian Amazonia has been the region most studied. Three anthropological studies in particular merit discussion: Wood and Schmink (1978), Schmink (1982), and Bunker (1985).

Wood and Schmink (1978) analyzed the shift in the Brazilian colonization program from the support of small peasants to the promotion of large-scale enterprises. The state blamed the small peasant for not being able to capitalize, produce for the market, and curtail "predatory" forest burning, while at the same time the state was not making good on its promise to provide infrastructure and financial and administrative support to overcome constraints on market access, and credit and tenure security during the early years of settlement. A brief documentation of the difficulties that a small farmer faced when claiming land is presented, calling attention to the circularities in the process.

Schmink (1982) later published an article concentrating solely on land conflicts in the region. She concluded that "conflicts over land in the Amazon is an expression of fundamental contradictions in Brazilian state and society" (p. 341). She identified two types of confrontation: one between native groups and members of the national society and one between *posseiros* (occupiers of small parcels of land) and owners of large estates. Her analysis is centered on the latter type of conflict, when land that is already occupied by posseiros is bought by an outsider. The purchaser tries to clear the land and develop mechanisms to convince posseiros to leave the area, such as direct violence and/or monetary offers for improvements. When the posseiro is not killed during direct "negotiations," the parties resort to government agencies. Under Brazilian legislation, the "social function of land" is recognized (i.e., it allows official agencies to expropriate land that is not socially productive, such as unoccupied land [p. 348]). However, a titled but absent claimant has precedence over a posseiro. Instead of filing for a title, posseiros prefer to sell land rights attained by "improving" the land (i.e., clearing the forest) and sanctioned by a Licencia de Ocupação (License of Occupation). The practice of selling these rights is illegal and has been denounced as an act of "bad faith" serious enough to make the posseiro undeserving of compensation for clearing the land, if and when a titled owner comes along.

Bunker (1985) tries to explain what he considers the underdevelopment of the Amazon: extraction takes precedence over production, leaving the region poorer as population is attracted. As part of his study, he documents in great detail the problems of small peasants when faced with the demands of modernization by the state: the need for a legal land title, commercialization of output, demand for credit, and technical assistance. His hypothesis is that the costs and benefits of adapting to "modern" institutions are not homogeneous for individuals. This brings about social inequality and a retardation of developmental goals (p. 151). When settling in frontier areas, there are several state agencies that farmers must deal with. A title or License of Occupation for the land must be procured if access to credit is desired. A License of Occupation could take some years to obtain and could secure only short-term credit for annual crops, even though sometimes the best use of the land would be perennials, which demand higher investments—for which credit could only be gained with a title. Technical assistance is obtained through approved credit. In 1977, securing credit required between 15 and 32 trips from the farm to the different agencies and took between 8 and 91 days to complete (p. 165). Commercialization was costly due to poor infrastructure, although more roads would attract more would-be claimants. Learning all the "proper" procedures may require years, enough time to cause the failure of farming in the Amazon.

In summary, these studies show how cumbersome legislation not backed by the resources necessary to enforce it affects the way peasants access and use productive resources. Unlike 19th century America, the Brazilian state seems to be more influential in this process of formation of property rights.

Institutional Constraints in Peruvian Amazonia

In this section, the process of formation of property rights in the Peruvian region of Ucayali is presented. Information about land conflicts was collected in the city of Pucallpa, capital of the Department of Ucayali (Figure 9.1), during the fall of 1991. The basic source of information was official files in the Ministry

Figure 9.1   Peru and the Department of Ucayali.

of Agriculture regarding conflicts over land tenure and titling to groups. Interviews were also conducted with officials responsible for resolving conflicts, dealing with titling issues, and granting land. The evidence presented here is based on case studies rather than random samples. Systematic evidence was not available. It is important to note that information was collected well before the current massive titling projects were put into effect.

In less developed countries, the budget available for people to execute the laws that affect resource allocation is usually very limited. To illustrate this point, the following section describes the institution responsible for giving and titling land and for dealing with land conflicts in the region. Detailed descriptions of the procedures required by law to obtain different kinds of land are contained in the second subsection, and two illustrative cases are described. Following that, five case studies on pending land conflicts are presented. They illustrate how people secure tenure and shape property rights in the region. Finally, this empirical evidence is contrasted with the preceding description and some conclusions are drawn.

La Secretaría de Asuntos Productivos Extractivos (SAPE)

When the empirical research for this project was done in 1991, there were still traces of the regionalization which began in 1990, when several state functions were decentralized. Under this scheme, each region was planned to have just six agencies (secretarías). All functions previously in the hands of offices of the Ministry of Agriculture were assigned to SAPE, but with the same personnel. Actions regarding agrarian reform (land redistribution and/or consolidation) were managed by Dirección de Reforma Agraria y Apoyo Empresarial (Office of Agrarian Reform and Entrepreneurship Support). When agrarian reform was decreed in 1969, this office was called the Office of Agrarian Reform and Rural Cadastre (a cadastre is an official registry of real estate used in apportioning taxes). The name change seems to respond to the general criticism that no managerial support was given after land was distributed.

In the region of Ucayali, three officials and two secretaries worked in the office responsible for allocating land. The director was an engineer, the official responsible for resolving conflicts was an agrarian technician, and the official responsible for allocating land to native communities was a high-school graduate. Needless to say, the office does not have a rural cadastre. Plot location is indicated by neighbors' names or any available landmark (a creek or a road).

What Is the Legal Procedure to Obtain a Piece of Land?

The standing forest, the land underneath, and the resources below the surface (e.g., petroleum) are all the state's property. There are two different procedures

available to private individuals regarding access to land for agricultural purposes: individuals can be given a title or Certificate of Possession, or they can
be given access to the land to cut down and market the timber. For native ethnic
groups, the procedures are slightly different and are described separately.

Agricultural Land

Anyone can occupy a piece of land and, after clearing it, write a petition (solicitud)
to SAPE. This petition initiates a process that is supposed to prove that the
claimant has had direct and peaceful control of the land for at least a year. Direct
control is proven by the so-called "improvements" (i.e., investments or work
done in the area). In the words of the official responsible, these are clearing
forests, planting crops, constructing buildings, etc. "Peaceful control" depends
on what your neighbors say about you. The initial petition costs less than one
dollar in official fees, on top of the cost of visiting Pucallpa, the regional capital.

After the petition is accepted, one *in situ* inspection must be carried out; the
costs are incurred by the claimant and depend on proximity to Pucallpa. To
schedule the inspection, the claimant has to go to the Pucallpa offices of SAPE;
otherwise, the file may get lost. The *in situ* inspection has to be publicly an-
nounced by notices at the district municipality where the land is located, at the
provincial municipality, at SAPE in Pucallpa, at the actual claimed land, and at
the local political authority (teniente gobernador) for at least five days, one week
prior to the visit.

After the visit, a technical report is issued. It states the total area claimed, the
area cleared, land use (crops), any buildings—all these are so-called "improvements"—plus any land left for fallow (purmas) and includes location drawings.

If no other claimant comes forward, a certificate of possession is granted by
the director of agriculture at SAPE, who in turn reports to the regional secretary.
For the years 1989–91, the number of titles granted in Ucayali was 2210, 2425,
and 621, respectively. The area awarded depends on the extension of the improvements and population pressure (the area is larger in less populated areas).

The Certificate of Possession is legally valid only to apply for loans at the
Agrarian Bank, but in practice people buy and sell "land improvements" backed
by this certificate, as discussed later.

The granting of a title requires some extra steps. To begin with, an extra
technical visit and a legal opinion by the lawyer at SAPE are required. The
added *in situ* inspection will locate the plot with geodesical information and
classify the land according to its potential. The legal opinion is supposed to back
the claim with relevant legislation and to clear the claimant's land from other
legal claims, although this is far from true. Formally, a certificate stating that the
land claimed is not already registered by somebody else is also required. This

is obtained from the Public Registry. If the land is already registered, the process of reversing possession from the registered owner to the state is started and follows the above-mentioned steps (petition, *in situ* inspection, legal opinion, resolution). The registered owner is not notified.

Titles are signed by the regional secretary. Most titling files were for groups that began the process themselves and were finally given free adjudication contracts (contratos de adjudicación a título gratuito) with the following conditions:

- Do not abandon the parcel for more than two consecutive years.
- Do not allow a third party to partially or totally occupy the land.
- Do not sell—partially or totally—your rights over the land.
- Start exploitation within the first 12 months of signing the contract.
- Do not cultivate coca, but follow SAPE directions for crop substitution.

In short, you cannot freely and legally dispose of the land being given to you for free for at least three years. A thorough *in situ* inspection is then done. Each would-be owner is listed, and family demographics are collected. All information is provided by the potential beneficiary. Incentives are therefore in place to overstate land clearing, land use, and "improvements" in general because the land area awarded depends on, among other things, the area with improvements. Two cases are briefly reviewed as examples:

1. Parcelación Neshuya-Curimaná, located close to the Von Humboldt National Forest, at km 60 on the Federico Basadre Highway (which connects Pucallpa to Lima; see map) and 13 km adentro (into the forest, accessible by a poor road). The group of 51 potential beneficiaries initiated the titling process. They are mostly male, range in age from 26 to 74 years (the median being 37), and are relatively highly educated, including some college education. Almost all have banana trees and manioc crops (banana being the commercial crop), some have rice, and very few have citrus trees. Areas for each crop are listed as no bigger than 1 ha, while 20 ha was given to each person. All plots were designed to face the road to Curimaná. The process of adjudication began in 1988 and titles were finally ready in 1990.

2. Parcelación El Ensueño, located at km 50, 16 km adentro. This also borders the Von Humboldt National Forest. The process was initiated by the people themselves, four women and two men, who were given extensions ranging from 18 to 52 ha. Their median age was 42, the youngest being the most educated ones. They have banana and corn, with manioc as the third most important crop. Although the technical report classified the land as apt for grazing, titles were given anyway. The process began in 1988 and was finished in 1990.

In October 1982, Jesus Gonzales Ruiz petitioned for a title over land for which he had a Certificate of Possession. The plot is located at km 8.6 and is 15 ha in size. The *in situ* inspection was made in July 1983, confirming the existence of manioc, bananas, and citrus trees; one reforested hectare with bolaina, cedro (cedar), and caoba (mahogany); and 480 seedlings of different tree species. Wire fences, a house, a hut, and a pigsty were also seen. Gonzales also presented a project to raise pigs. In 1991, the Resolution of Adjudication was ready, giving him 7 ha. Because it included a commercial project, Gonzales had to pay a total of $15 over 20 years for the land, valuation that resulted from the Arancel de Areas Rústicas (Legal Guidelines on Rural Land Prices). In August 1991, Gonzales paid the whole amount and received his title, with the same restrictive conditions mentioned before.

If some piece of land is already occupied and the land's potential classification does not designate it to be apt for agriculture, a Certificate of Possession or a title may be issued anyway, unless the claimed land is located on a classified national forest. In the latter case, possession is not recognized, but "cesión en uso" (a type of concession) is issued, which implies that the main use is not agricultural but forestry, although in fact people allocate the land to agricultural uses after clearing it. This process has led to new regulations regarding land in national forests, actually reducing the national forest's size by 5 km on each side.

Forest Land

If an individual occupies land for agricultural purposes, clears the standing forest, and either burns it or builds his house with it, no compensation is paid to the state, which is the legal owner of the trees. However, if an individual wants to sell the timber, he must obtain a permit from the Office of Forestry and Environment of SAPE. The Office of Forestry and Environment does not have a final version of a rural cadastre, although it has been working for a year to assemble one.

There are two kinds of authorizations, depending on the classification of land that an individual plans to clear. For agriculturally suitable land, one has to obtain a permit. For land located in free access forest, one has to obtain a contract.

To obtain a contract, a person must locate a piece of forest, draw up a rough sketch, make a study of varieties of timber (at least six) and available volume, and write up a petition. No *in situ* inspection is required since the officials trust the draft of a cadastre that they have. Officials say that the whole process takes a week in their offices. When approved, the contract is valid for two years if the size of the area is less than 1000 ha. For larger areas, a map at 1/20,000 scale, the identification of at least 20 species, plus a thorough economic feasibility study are required. In exchange, the contract can be valid for at least 10 years and at most 20. The contractor pays 5% of the 50% estimated value of the species and volume that he plans to cut. This value is of the standing forest and is determined by SAPE. If the contractor does not want to reforest (the new trees he plants are state property and consequently he does not have any right over them), he has to pay what is called a "canon" to help finance reforestation efforts by the office.

A permit is given for land suited for agriculture. The person must show either his title or the Certificate of Possession and also an estimate of the volume and value of the species he plans to extract. No canon or guarantee fee is required.

Each piece of timber that the contractor extracts should have written on it the contract number so that the Forest Police can check the validity of the contract. However, around Pucallpa there is only one checkpoint, which used to be closed from 6 p.m. to 8 a.m. Sometimes the Forest Police check the sawmills, and the contractor will be held responsible for any penalties if any irregular pieces (i.e., pieces without a contract or permit number written on them) are found.

Land to Ethnic Groups

The granting of land to ethnic groups was important in Perú in the 1970s and 1980s, pushed by agrarian reform. Ethnic groups are subdivided in different villages, officially called native communities. About 130 communities received official recognition and titled land in that period in Ucayali alone, out of a total of about 300 communities.

The procedure consists of two steps. First, the community must be recognized (tener personería jurídica), and second, a land title is issued. To be recognized, a communal census must be conducted, from which a socio-economic diagnosis is made. Both a technical and a legal opinion are required. If these opinions are favorable, the community is inscribed in the Official Community Register, administered by the regional secretary.

The census and the socio-economic diagnosis must be paid by the interested party. Needless to say, the level of monetization of the economy of these groups is very low, and they usually do not see the necessity to be legally recognized. The initiative and payments toward recognition are generally undertaken by local private advocacy groups—financed by foreign foundations—and also by the Catholic church. How quickly the process is completed depends on the political will of the party in power.

After legal recognition is obtained, the procedure to title agricultural land is initiated. Delays can be experienced at several points in the process. According to the president of the most important ethnic group in the region, the regional secretary refuses to sign the titles because he is aligned with the timber extractors.

If a community wants to sell timber, a permit must be procured, following the procedures to title forested land. Although communities are not required to pay taxes, they must pay for the permit to extract timber. Because they usually do not have the means to pay for this permit, they retain the standing forest. Unfortunately, forest extractors, backed by licenses, cut down the forest (remember that SAPE has at most only a sketch of a cadastre and no *in situ* inspection is required to grant a permit or contract). Bargaining between the community and the forest extractor revolves around compensating the community for the lost timber.

Conflicts Over Land: Case Studies

If we agree that conflict (i.e., the existence of overlapping claims) reflects tenure insecurity, then land tenure is most insecure when:

- Land is close to access roads or main highways
- The owner or posseiro is not continuously present (one is advised live on the land)
- The owner is not present and does not get along with his neighbors
- "Improvements" have not been made; even if some portion of land is left for fallow (purma), the owner may lose it

A brief summary of some cases illustrates these conclusions.

1. *Villa María Inez*—This 100-ha property is not directly administered by the titled owner; he appointed a guardian. This is error number one because under Peruvian law, absentee titled owners lose their rights. It is located at km 10.5 on the highway that begins in Pucallpa and connects to Lima. SAPE recognizes the previous owner instead of the current titled owner. According to the current owner, the former director of agriculture at SAPE organized an invasion with other officials, rapidly distributed Certificates of Possession, and began reverting dominion from the owner to the state. Mestanza, who obtained his Certificate of Possession for 3 ha in less than one month, was challenged in his claim by Pelaez. In March 1991, Pelaez claimed that the 3 ha given to Mestanza were his; he had bought the "improvements" (made with a tractor) three years earlier for about $1000 but had become ill and had to abandon the field. An *in situ* inspection was made, with Mestanza's neighbors as witnesses of his work, and supposedly an agreement was reached and signed in which Pelaez conceded to be compensated monetarily for the land awarded to

Mestanza. However, Pelaez showed up at the SAPE office, claiming that his signature was forged and that he was not present when the agreement was reached.

2. **Fundo Neno**—This 40-ha area is located at km 25 on the highway. One Certificate of Possession was issued in July 1990 to Chávez and another to Cárdenas in February 1991. Then, Vásquez, a third party, asked to nullify the certificates. The valid certificate is supposed to be the oldest one. Chávez had asked for his certificate in 1986. Cárdenas was the guardian. The local political authority (teniente gobernador) backed Cárdenas with a document claiming that Cárdenas had lived in the area for more than four years. The *in situ* inspection in August 1991 found that Fundo Neno was divided in two: one part bought by Vásquez in December 1990, on which he has cattle and has also cleared 2 ha, and another part occupied by Cárdenas, on which 2 ha has been cleared. Both individuals have built houses (an important sign of occupancy). The technical report by SAPE, which confirmed the validity of Chávez's certificate, recommended that Cárdenas' certificate be nullified, and that Vásquez be told to stop working on the field.

3. **Fundo Rosanita o El Valiente**—This area is located at km 6 and 3 km adentro from the highway. In 1988, the occupant for the owner, Rosanita, asked for adjudicación gratuita (free adjudication) over land on which she made improvements in 1987. In 1989, the *in situ* inspection confirmed the existence of several crops (manioc, banana, citrus, mango) plus fallowed land and standing forest, covering 5 ha. In July 1989, Rosanita complained of damages caused by Tulumba, who allegedly cut producing trees. Tulumba asked for his Certificate of Possession in December 1989, saying that he had cleared 2 ha and planted bananas, corn, manioc, rice, and fruit trees. He claimed that Rosanita's occupant had cut his other crops, some of which were 15 years old. Another *in situ* inspection in 1990 confirmed that Tulumba had more crops and trees than Rosanita's occupant and suggested dividing the land proportionally. In August 1990, another solution was suggested: that Tulumba leave the area. In February 1991, a different report recommended that Rosanita's possession be nullified and the land awarded to Tulumba!

4. **INIIA vs. Noriega**—In February 1989, INIIA, a public research institution, asked for what it termed abandoned land at km 26, 3.5 km adentro, to set up an experimental station with Japanese cooperation. In April, the *in situ* inspection report asserted that the solicited parcel was part of a subdivision of which 35 ha was given to Rengifo, who transferred the land to Noriega. According to the report, Noriega had not worked the land for four years. There was an abandoned house, five-year-old fruit

trees, and old purmas. In May 1989, a Certificate of Possession was given to INIIA. In July, Noriega asked to have his land back and accused INIIA of invading. He showed a Certificate of Possession from 1986 (given just one month after he bought the "improvements"). Another *in situ* inspection was scheduled for August, during which a neighbor said that Noriega had not been there for the past year. A statement from Noriega asserted that he had had an accident and needed hospital care. This left INIIA's guardian to take care of the plot in exchange for corn planted. The guardian said that there was no corn. Noriega said that he was robbed by the guardian. To make matters worse, a fire in October 1989 consumed the offices of SAPE. In October 1990, another *in situ* inspection was scheduled but not carried out.

5. *Varzeas*—Sifuentes had been a posesionario of a 10-ha lot since 1970 and found Saavedra dividing the land and usurping SAPE functions. Sifuentes claimed that he was unable to plant crops in 1990 because of well-known credit shortages at the Agrarian Bank and because his tractor had broken down. He held a Certificate of Possession from 1988. His land borders with Ucayali River, one neighbor, and floodable land, over which no Certificate of Possession could be granted. Saavedra asked for an *in situ* inspection. He planned to divide 18 ha among 18 people. The *in situ* inspection report said that there was corn, a small hut, and some recent forest clearing. Sifuentes claimed that he could not sue Saavedra because it was too costly. He showed another Certificate of Possession dating 1989 and loan contracts from the Agrarian Bank. SAPE's solution was to allow Saavedra to subdivide and to reduce Sifuentes' tenure to 5 ha. During the rainy season, the area in conflict disappears due to floods. Legally, then, Sifuentes should not have had a Certificate of Possession in the first place, according to the law governing land occupation in Peruvian Amazonia. Another *in situ* inspection accused Sifuentes of being conflictive and employed the police to keep order. Both the Agrarian Bank and the regional political authority (prefecto) sent letters to SAPE backing Sifuentes.

Some points should be noted. First, dates are written at the officials' will, usually biased toward showing faster completion of procedures. Second, the files that were reviewed are part of the first piece of a process that can take a long time. If one of the parties is not satisfied with SAPE's ruling, which is the administrative stance, then the complaint can be taken to the agrarian jurisdiction (fuero agrario), which means lawyer fees and court costs. Moreover, because Pucallpa does not have a fuero's branch, litigation must be taken to Huanuco, about 375 km from Pucallpa. Getting there takes approximately 10 to 12 hours by bus.

Conclusions and Policy Recommendations

Overlapping land claims suggest an increased perception of land scarcity and, therefore, devoting resources to gain exclusive access. In effect, land that can be profitably dedicated to produce commercial goods should be easily accessible to the marketplace. This land is close to roads; however, the extension of roads in the Ucayali region has neither increased nor improved recently, while people have continued to settle in the area. The proposition that land in Amazonia is increasingly scarce because of population pressure may sound counterintuitive, but the evidence seems to support it.

Under the Peruvian legal framework in effect at the time this research was conducted, tenure security is attained by making land-specific investments to prove occupation by actually living in the claimed plot. In Peruvian Amazonia, the law forces settlers to change land use from primary forest to agriculture in order to claim tenure. In this sense, Peruvian law is very similar to the provisions of the Homestead Act in the United States in the 19th century and very different from a situation of no legal incentives for occupation (which would merely result in a spontaneous frontier). It is therefore reasonable to conclude that regulations regarding land access are distorting people's decisions about occupying Amazonia and converting its land to new uses.

Productive, direct occupation, which is equivalent to clearing the land, planting crops, building fences and houses, and living on the plot or very close to it, is more important than having a title. Since occupation is primary evidence, a title held by an absentee owner becomes secondary. If land conflicts arise, the person who "improved" the land is at least entitled to compensation, the amount of which is decided at SAPE by direct negotiation between the parties.

Resource degradation, land-specific investments, and physical presence signal occupation and are the means to attain tenure security. The proof of direct occupation is both the so-called improvements (clearing the forest and planting crops) and recognition from neighbors. This requirement is important in preventing the use of violence to settle conflicts. Moreover, conflicts usually arise among small farmers, for whom it is reasonable to assume that the ability to use violence is relatively homogeneous. The latter would also justify a spontaneous allocation of relatively homogeneous land sizes, which is what we observe, but through the mediation of a bureaucratic agency.

Titles are a secondary component in gaining exclusive access, as reflected by the estimated number of Certificates of Possession and titles granted per year. Moreover, the costs that the farmer must bear to obtain a Certificate of Possession are substantially lower than filing for title. Even if one lives on one's own titled plot and someone comes along and clears a part without the owner noticing, that person can obtain a Certificate of Possession. Land use mostly involves a very short-term planning horizon: annual crops predominate. However, fruit trees are also planted as a way to signal occupation, since they can be dated and thus support a land claim.

The lack of resources at the office responsible for granting land plays a role in tenure insecurity. Inspections are scheduled according to the claimant's interests; the public office does not have a map and cannot hire such services as topography. Access to local power plays a role in that "letters of recommendation" are written backing particular claims. Rents are dissipated at the private level. Each inspection is paid for by the claimant; every resolution at SAPE should be pursued, by the claimant in person and even bribes to officials. Forest clearing is necessary to prove a claim, and the condition to market timber (i.e., to have a specific permit) is to have a Certificate of Possession. One cannot get a Certificate of Possession unless one has cleared, burned trees, and planted something. Moreover, a posesionario can burn virgin forest and pay nothing in compensation, while marketing timber requires permits and resources to be consumed in the process of obtaining them. There seems to be a resource transfer from claimants to bureaucrats through the payment of inspections and possibly bribes and a resource transfer from the state to claimants in the form of burning virgin forest and occupying land.

This chapter has illustrated the institutional constraints to assigning fragile lands to their most valued use, taking into consideration all the non-marketed benefits that tropical forests provide to society. Existing constraints seem to have been justified on the grounds of respecting land zonification and guaranteeing land access to the poor. However, they achieve the opposite objectives and serve to secure rents to public servants. When designing regulations to gain land access, policy-makers should examine the consequences and consider all the possible effects these procedures might have.

Considering the benefits, or "positive externalities," that Amazonia provides to the whole planet, action must be taken. First, regulations regarding land access and use should correctly weigh the social benefits of the migration of people away from cities against a variety of benefits resulting from the conservation and ecologically sustainable exploitation of forest resources. Second, some form of international coordination effort should be undertaken so that reduced monetary benefits from missed development opportunities in Amazonia are compensated by some form of debt reduction, increase in foreign loans, or investment at low interest rates. Third, the recognition of indigenous rights is key to longer term management of natural resources in Amazonia, and this includes international treaties and intellectual rights. Last but not least, information should be disseminated around the world about the global benefits of natural resource management in Amazonia, as well as the costs and which countries are bearing them. The mere existence of the South American rain forest has values that may well exceed the monetary benefits of development.

References

Alchian, A. and Demsetz, H. (1973) The property rights paradigm. *Journal of Economic History,* 23(1): 16–27.

Alston, L., Libecap, G., and Schneider, R. (1991) *Property Rights, Rent Dissipation, and Environmental Degradation in the Brazilian Amazon.* University of Illinois, Urbana-Champaign, IL.

Amemiya, T. (1981) Qualitative response models: a survey. *Journal of Economic Literature,* 19: 1483–1536.

Anderson, T. and Hill, P.J. (1975) The evolution of property rights: a study of the American West. *Journal of Law and Economics,* 18(1): 163–179.

Anderson, T. and Hill, P.J. (1983) Privatizing the commons: an improvement? *Southern Economic Journal,* 50: 438–450.

Anderson, T. and Hill, P.J. (1990) The race for property rights. *Journal of Law and Economics,* 33: 177–197.

Barrantes, R. (1992) Land Tenure Security and Resource Use in Peruvian Amazonia: A Case Study of the Ucayali Region. Ph.D. dissertation. Department of Economics, University of Illinois, Urbana-Champaign, IL.

Bunker, S. (1985) *Underdeveloping the Amazon: Extraction, Unequal Exchange, and the Failure of the Modern State.* University of Chicago Press, Chicago.

Carter, M., Wiebe, K.D., and Blarel, B. (1991) Tenure Security for Whom? An Econometric Analysis of the Differential Impacts of Land Policy in Kenya. Department of Agricultural Economics, University of Wisconsin, Madison, WI.

Ciriacy-Wantrup, S.V. and Bishop, R.C. (1975) Common property as a concept in natural resource policy. *Natural Resource Journal,* 15(4): 713–727.

Cotlear, D. (1989), Desarrollo Campesino en los Andes. Report of the Instituto de Estudios Peruanos, Lima, Perú.

Demsetz, H. (1967) Toward a theory of property rights. *American Economic Review,* 57(2): 347–359.

Feder, G. and Feeny, D. (1991) Land tenure and property rights: theory and implications for development policy. *World Bank Economic Review,* 5(1): 135–153.

Feder, G., Onchan, T., Chalamwong, Y., and Hongladarom, C. (1988) *Land Policies and Land Productivity in Thailand.* Johns Hopkins University Press, Baltimore, MD, for the World Bank, Washington, D.C.

Feeny, D. (1982) *The Political Economy of Productivity.* University of British Columbia Press, Vancouver, Canada.

INE (1986) *ENAHR, Encuesta Nacional de Hogares Rurales, Resultados Definitivos.* Instituto Nacional de Estadística, Lima, Perú.

Johnson, O. (1972) Economic analysis, the legal framework and land tenure systems. *Journal of Law and Economics,* 15(1): 259–276.

Kennedy, P. (1985) *A Guide to Econometrics,* second edition. MIT Press, Cambridge, MA.

Libecap, G. (1989) *Contracting for Property Rights.* Cambridge University Press, Cambridge, U.K.

Maddala, G.S. (1983) *Limited-Dependent and Qualitative Variables in Econometrics.* Cambridge University Press, Cambridge, U.K.

Repetto, R. (1989) Economic incentives for sustainable production. In: *Environmental Management and Economic Development.* G. Schramm and J.J. Warford, Eds. Johns Hopkins University Press, Baltimore, MD, for the World Bank, Washington, D.C., pp. 69–86.

Schmink, M. (1982) Land conflicts in Amazonia. *American Ethnologist,* 9: 341–357.

Somerwitz, H.S. and Fahr, S.M. (1967) The Peruvian Land Registration System and Some Suggestions for Its Improvement. Iowa Universities Mission to Perú, in cooperation with the Agency for International Development, Lima, Perú.

Southey, C. (1978) The stapes thesis, common property, and homesteading. *Canadian Journal of Economics,* XI(3): 547–558.

Southgate, D. (1990) The causes of land degradation along "spontaneously" expanding agricultural frontiers in the Third World. *Land Economics,* 66(1): 93–101.

Stroup, R.L. (1988) Buying misery with federal land. *Public Choice,* 57: 69–77.

Umbeck, J. (1977) The California gold rush: a study of emerging property rights. *Explorations in Economic History,* 14(2): 197–206.

Umbeck, J. (1981) Might makes rights: a theory of the formation and distribution of property rights. *Economic Inquiry,* 19(1): 38–59.

White, K.S., Wong, D., Whistler, D., and Haun, S.A. (1990) Shazam. Econometrics Computer Program, McGraw-Hill, New York, NY.

Wood, C. II and Schmink, M. (1978) Blaming the Victim: Small Farmer Production in an Amazon Colonization Project. *Studies in Third World Societies,* Publication Number 7, 77–94.

Property Rights, Nature Conservation, and Land Reform in South Africa

10

Neil Adger

Centre for Social and Economic Research on the Global Environment, University of East Anglia, Norwich, U.K.

Abstract

Land reform is a key element of any transition to a sustainable and equitable development path for South Africa. The arguments for and against large-scale redistribution of land center on the issues of agricultural productivity and utilization, on equity and justice issues, and on the ecological and environmental consequences of alternative land uses. In areas presently used for nature conservation, evidence is often presented that nature conservation activities, particularly with tourism revenue, can be the most "profitable"; therefore no redistribution is desirable on efficiency grounds. This chapter argues that categorizing land uses such as agriculture and nature conservation as mutually exclusive is unhelpful and outdated, as demonstrated elsewhere in Africa. In order to integrate nature conservation with other land uses, redistribution of property rights allowing multiple land uses is required. These issues are reviewed with reference to a case study of the northern Transvaal region.

Introduction

Since 1994, South Africa has been engaged in a reconstruction and development program to set the country on a path to sustainable and equitable development,

1-57444-077-2/97/$0.00+$.50
© 1997 by CRC Press LLC

overturning the decades when the state apparatus and the resources of the country were used for the benefit of a minority at the expense of the majority. Since 1994, South Africa has had high expectations of equality of opportunity and of resource use. Conflicts over control and access to resources, though no less acute, are now tempered by the legitimacy of a government of national unity. The objectives of sustainable and equitable development are necessarily constrained by political and economic imperatives and by the structure of the economy inherited from the previous government.

Nature conservation and habitat protection are undoubtedly important components of resource sustainability in South Africa, along with equitable distribution of both wealth and income. This issue is manifest in the debate over land reform. Disputed land in particular instances overlaps with present uses of land for nature conservation. This highlights a dilemma, though, as this chapter argues: alternative land uses and ownership do not form exclusive trade-offs between agriculture and nature conservation. It is important at the outset to distinguish between, on the one hand, land reform and redistribution, which essentially concern ownership, and, on the other, the property rights attached to all resources. Land redistribution has been a key agricultural and economic policy issue throughout the world, occurring for diverse reasons, often in association with radical changes in government and political structure. Property rights is a broad concept referring to bundles of rights and responsibilities associated with resources. The rights may be to use resources or to have access to them and can be transferable or non-transferable. Hence, property rights are much broader than simply tenure rights or freehold of land.

Land reform and redistribution of land associated with nature conservation, particularly private nature conservation, have been resisted in South Africa through appeal to the greatest "productive" use of the land. This is the same argument that was previously used in the redistribution of agricultural land in such newly independent states as Zimbabwe: that land should be redistributed as private farms, as these have the greatest productivity and are therefore of greatest benefit to the national economy. Similarly, in the sphere of nature conservation, it is often argued that retention of private nature conservation is the most productive use of land when the functions of biodiversity conservation and the revenue from nature tourism are incorporated.

However, these arguments ignore the equity considerations involved in redistribution. They also accept the necessity of productivist private agriculture or nature conservation with high financial returns rather than promoting communal ownership and control. Private ownership is a prerequisite of neither increased productivity nor sustainable land use practices. Further, advocacy of the status quo promotes a single concept of nature conservation and biodiversity based on tourism interests and a myth of wild Africa, rather than any assessment of multiple land use. The chapter concludes that rights to resource use and to the economic benefits from existing nature conservation areas are necessary to integrate conservation with development.

Property Rights and Sustainable Development

Economists have long recognized that there may be many reasons why the environment is not used in a sustainable manner. The underlying reason is that the environment is treated as a free good because of its public good nature. A public good is one whose use by others cannot be restricted; it is non-excludable and non-consumptive. None can be excluded from its use, even if they do not pay for it. Second, one person's use of a pure public good does not detract from anyone else's use of it.

However, it has been demonstrated that the proximate causes of environmental degradation are numerous and do not have any simple relationship with who owns or controls resources. Environmental degradation can take place where ownership rights to land are private, where they are communal, where they are controlled by the state, or when resources are essentially open access. Open-access resources are clearly liable to non-sustainable utilization, although Bromley (1989) points out that many analyses confuse all non-private property systems with open access (e.g., in Hardin's [1968] hypothesis of the Tragedy of the Commons).

Environmental degradation can take the form of *resource depletion* or *resource degradation,* which decreases the ability to maintain economic activity in the long run. Degradation of renewable soil and water resources and subsequent environmental stress in low-income, agriculture-dependent rural areas are often regarded as the major environmental issues in sub-Saharan Africa, although both the scale of the issues and the paradigms of causation of environmental change are contested. Environmental stress can be defined as the set of forces impacting on the natural resource base from which individuals sustain their livelihoods and derive economic benefits. These forces may be internal to the use of the resources or external political, economic, or environmental phenomena. For example, local environmental changes with negative impacts, such as loss of soil fertility, are not necessarily due to local action but may have external causes. For example, resource degradation may be exacerbated by ecosystem stress associated with global environmental changes in climate or other phenomena (see Hulme and Kelly [1993] on desertification, for example).

The key issue, highlighted by the South African case, is that natural resource degradation is brought about by the interaction of environmental with institutional, political, and economic factors. For example, factors external to the land-use system can come from inward investment in natural resource industries or from a change in terms of trade or institutional structure which encourages land use leading to negative environmental consequences. The internal factors caus-ing environmental degradation include migration and demographic change. The key mechanism by which these factors cause environmental degradation has been hypothesized as the breakdown in local institutional arrangements which enabled efficient and sustained use of resources (Kates and Haarmann, 1992).

External factors encouraging resource depletion and degradation can range from perverse national policies to international indebtedness, all of which may have direct or indirect impact on the natural resource base. South Africa has never been heavily indebted: the interest on its present external debt amounts to only 3.3% of export earnings (Overseas Development Institute, 1994). South Africa is thus not faced directly with a critical need for economic policy reform or the need to deplete its mineral and other subsoil resources due to such external pressures as indebtedness. However, internally, agricultural pricing policies at the national level encourage non-sustainable land uses. In South Africa, these policies have traditionally included capital grants and subsidies for many in-vestment activities, such as groundwater extraction and irrigation; purchase of agricultural machinery; subsidies on fertilizers and other inputs, including elec-tricity; and guaranteed producer prices. Commercial agriculture imposed costs on surrounding displaced populations by excluding access to water and other resources.

Thus, the South African case illustrates a range of environmental problems, facilitated by state intervention, which subsidizes land use to secure political influence in rural areas, in ways which Bromley (1994) has likened to state control in the agrarian sectors of the former Soviet Union. Nature conservation in South Africa has also been subject to policy interventions which have been unrelated to the goal of sustainable land use. Both state-designated protected areas and areas used for non-commercial private nature conservation uses ful-filled the dual roles of excluding local populations, as well as generating eco-nomic benefits for the state and private landowners. With the removal of the illegitimate purpose of control of rural populations and land, a reconsideration of agricultural policies and rural land use is taking place in South Africa, fueled by pressure for land redistribution (Bromley, 1995).

Land Reform in South Africa

Issues and Trade-Offs

This chapter focuses on the issue of redistribution of property rights away from sole-use nature conservation toward multiple land use or toward extensive ag-riculture. This is discussed in the context of evolving land reform policy in South Africa. In southern Africa, there is much experience with novel property rights regimes designed for the purpose of nature conservation. The CAMPFIRE program in Zimbabwe, for example, has often been held up as successful communal ownership and control of rights to wildlife (Child, 1993; Western and Wright, 1994). This scheme essentially creates communal rights to utilization or rights to sell wildlife resources on non-state land, where previously wildlife has in effect been an open-access resource. However, these schemes have not been without their critics with much debate between conservation biologists on the ethical and sustainability basis of these forms of utilization. For some, conservation of species through allowing the right to hunt for sport is too high a price to pay (Favre, 1993). Others argue that the commercialization of wildlife for consumptive uses is inherently unsustainable (Ehrenfeld, 1992). The practicalities of CAMPFIRE schemes, such as in the allocation of revenues and devolution of decision making within communities, are inevitably problematic (Mbanefo and de Boerr, 1993), because ultimate ownership of wildlife resources, in most cases, remains in the hands of the state (Gibson and Marks, 1995). Nevertheless, some schemes demonstrate the potential benefits of reconceiving rights to resources, and those examples where significant community benefits exist are those where "locals receive direct benefits and actively participate in decision-making about wildlife" (Gibson and Marks, 1995, p. 955).

Although changing rights to use of resources may bring about community benefits, if the institutional arrangements are effective, the question still remains as to ownership of land. Is redistribution of ownership environmentally more desirable than the present situation? Redistribution may be superior in terms of social equity (in restoring previous land rights) and potentially allows innovative property rights and multiple resource use. Private and state nature conservation have easily demonstrable economic benefits but ambiguous nature conservation benefits which are difficult to determine.

The trade-offs in economy, equity, and ecology are not straightforward, as there are other considerations and contexts, such as the belief system and the symbolism attached to colonial-style nature conservation. The current wisdom in southern African nature conservation is based on interventionist management of wildlife resources and on perceptions that in general people should be excluded from protected areas set aside for nature conservation. These diverse issues are considered after discussion of the mechanisms and rationale for land redistribution.

Redistribution for Productive Agricultural Utilization

The land reform question in South Africa mirrors in many respects the redistribution question as already attempted in other African countries. The prime recent parallel situation is that of Zimbabwe. The similarities between Zimbabwe and South Africa are primarily in a publicly stated objective of reform, namely, the increased efficiency of the use of land ("judicious utilization of land a national asset," according to the South African Land Settlement Act, 1993). In other words, the prime consideration has tended to be the overall level of agricultural production rather than the redistribution of wealth per se (Cliffe, 1994). In the first five years of Zimbabwean independence after 1980, 35,000 households (approximately 250,000 people) were resettled on land formerly in the largely white-owned commercial agricultural sector. This was well short of targets for land redistribution, and subsequent targets have been revised at intervals since then. A proportion of the government purchase came from the primarily white-owned large-scale commercial farming sector, principally utilized for arable crops. A further amount of land purchase took place on land in the less favored areas suitable only for grazing. The major redistribution in Zimbabwe happened immediately after independence in 1980, thus, a significant proportion of the land had been abandoned by white farmers during the war or by others who were keen to sell.

Land redistribution did not occur on the scale desired by the newly independent government in Zimbabwe because the property rights of landowners were guaranteed in the negotiated constitution, a situation which no longer holds subsequent to Zimbabwe's new land act. In Zimbabwe, a large number of white-owned farms were available to be sold, through out-migration of the owners immediately after independence in 1980, or were already being squatted. This amounted to 2.5 million ha, with some 250,000 people settled. This situation is not mirrored in South Africa, where there are not large areas of land immediately available in either the private or state sector, despite the optimistic assumptions of the initial program of the Government of National Unity.

The Reconstruction and Development Program of the ANC and subsequently of the coalition government of national unity in South Africa in the period immediately following 1994 involved targets for large-scale transfer of productive agricultural land to small-scale producers. It stresses the productive utilization of redistributed land, proposing grants and soft loans (at less than the prevailing market rate) for purchase of this land for agricultural purposes. In October 1993, a major conference on land redistribution under the incoming government set out this strategy which, in essence, sent all political parties down the road of "willing buyer–willing seller" land purchase (see Williams, 1996; Deininger and Binswanger, 1995; papers in *World Development*, September 1993). This is a market-based solution, but with heavy government intervention to ensure fair prices for land.

Opposition to major reform in Zimbabwe focused on the apparent contradiction between environmental conservation and improving agricultural production. The first reason for opposition to land reform was couched in terms of the environmental degradation resulting from intensive use of the communal land areas by small-scale farmers.* This ignored, in the main, the evidence that small-scale farming does not result in greater economic losses due to soil erosion than large-scale farming and, importantly, that black farmers had until independence been concentrated on communal land of marginal agricultural quality. In reaction to earlier analysis that farming systems in communal areas cause large-scale environmental degradation, the analyses of Grohs (1994) and others have shown that the observed levels of soil erosion on *arable land* in the communal lands of Zimbabwe do not significantly affect productivity. Further, there has been reconsideration of the grazing regimes on rangeland in much of southern Africa, showing that present practices have insignificant impacts on economic productivity (Behnke and Scoones, 1992, for example).

The second argument, on food production, focused on the economies of scale in large-scale farming, concluding that redistribution would decrease overall production and hence national food security. Again, according to Cliffe (1994) and Lipton and Lipton (1994) among others, the evidence from Zimbabwe after a decade of land reform does not support this argument despite the impacts of severe drought in the early 1990s. In South Africa, the approach of the new government to ownership of redistributed land comprises an allocation of residential and productive agricultural land to those presently living in the township on the edge of metropolitan areas (comprising possibly 40% of the black population) and redistribution to those who have specific land claims from previous displacement. Overall, agriculture accounts for between 4 and 7% of gross domestic product, the large range being due to the sensitivity to climatic variations of commercial agriculture. Yet studies carried out in the period before the 1994 elections by such international agencies as the World Bank point to the inefficiency of the predominantly white-owned commercial sector (Overseas Development Institute, 1994).

Much of the land that is under state-owned or privately owned nature conservation use is in semi-arid areas and may not have the potential to support a dense population involved in intensive agricultural production. In the Transvaal, in a reaction to calls for "agriculturalization" of some of this land, Kruger National Park authorities and the South African conservation lobbies have been pushing for an expansion of the park to a contiguous protected area with other parks in Mozambique and with Gonarezhou National Park in Zimbabwe, and

---

* Similar analysis of the situation in Namibia by Pankhurst (1995) shows that the same arguments are presently being deployed in Namibia by interested parties, but that "the case against a major land reform is overstated and inaccurate" (p. 551) and that the status quo (of agricultural land use) "is itself thought to be intensifying environmental degradation" (p. 558).

even to expand into the private reserves in the Kruger vicinity. This plan also has international backing from the World Bank, with support for the infrastructure developments in Mozambique (Ellis, 1994).

The area surrounding Kruger National Park was also politically sensitive in the 1980s and early 1990s, with allegations that the South African Defense Force used the area as a base to assist the RENAMO rebels in Mozambique (see Ellis, 1994). The environmental policy statement of the ANC prior to the elections called for restructuring of Kruger National Park management and boards (*London Guardian,* April 29, 1994), although this has not been acted upon since the installation of the government of national unity.

Studies throughout South Africa show that local attitudes toward areas designated for nature conservation are, in the main, negative due to the perception that these areas uphold the segregated land-use system of the apartheid era and also because of the critical scarcity of agricultural land (see Hackel [1993] on Swaziland). In the Transvaal area, the conflicts are stark between nature conservation, both on state and on private land, and the need for economic development principally through agriculture.

The Principle of Conservation as an Exclusive Land Use

The history of protected areas in Africa shows that the guiding principles of conservation have been imposed by European images of correct "management," of their ideas of pristine "wilderness" existing before European colonization, and of assumptions of the relationship between African people and their local environment, including wildlife. The perceptions are based on little or no knowledge of common property regimes and management of resources by local people. These guiding principles have also, according to Adams and McShane (1992), been uniformly applied throughout Africa, thus ignoring the diversity of situations within the continent.

Management of wildlife in Africa has relied almost exclusively in this century on designating protected areas and excluding resource extraction and habitation within their borders. This system was formalized in the early 20th century due to perceptions of declining numbers of large mammals in some parts of colonial Africa. Agreement between the colonial powers, through two international conventions on conservation in 1900 and 1933, contributed to the establishment of parks. Kruger National Park was designated in 1926 and served as a model for many other areas in Africa. For 30 years prior to 1926, the area had been set aside as a hunting reserve simply to increase the stock of wildlife available for hunting by white hunters (see Carruthers, 1993). Local populations had already been banned from hunting for many decades. According to MacKenzie (1988), the first game legislation was introduced in the Cape Colonies by the Dutch East India Company in 1657.

In the 19th century, the portrayal of traditional hunting practices and range management by local people as detrimental to the environment and the cause of the loss of wildlife allowed these practices to be banned as part of colonial government "civilization" policies. According to Adams and McShane (1992), the evidence for the causes of periodic regional crashes in wildlife populations shows that such cycles may have had natural causes, such as climatic shifts or changes in cattle grazing due to tsetse infestation, and that hunting by colonials rather than local utilization had had more profound consequences. Despite this, the famous hunter Frederick Selous claimed in the early 20th century that of every 1000 hunted elephants, 997 were killed by Africans (Adams and McShane, 1992). Although the scale of commercial sport hunting is now a small proportion of that in the early 20th century and does not constitute a threat to wildlife numbers, the principles of exclusion in conservation remain.

Thus, the guiding principles behind nature conservation in Africa, as espoused in South Africa in particular, have stressed the need to minimize the contact of wildlife with African populations and more recently to stress the role of conservationists and hunters as the guardians of the threatened biodiversity. Such circumstances do not encourage innovative management or the instituting of novel property rights for integrated nature conservation and development, as attempted elsewhere (Stocking and Perkin, 1992; Wells and Brandon, 1992). The issues of resource utilization and the related issue of land ownership are now discussed in the context of the northern Transvaal region.

Transvaal Case Study

Background

As outlined previously, competing interests in the reformed South Africa face critical decisions on future development of land, particularly with regard to trade-offs between equity, efficiency (or productivity), and sustainability. This section presents details of these arguments for the northern Transvaal region of South Africa, where, as outlined above, there are both state- and privately owned areas currently used for nature conservation and related uses, where intensive agricultural development may be limited by the physical constraints of climate and environment, and where previous land-use policy has included displacement of human populations from areas under agriculture.

This displacement occurred in parallel with the creation of "non-independent homelands." In the northern Transvaal, one such homeland was Gazankulu, with an area of only 122,000 ha and a population density of between 150 and 300 people per square kilometer, compared to 19 per square kilometer in the surrounding largely white-owned area and possibly 30 per square kilometer nationally. Thus, high population density and low levels of agricultural infrastructure exist in such areas with only 3000 ha of irrigated farmland in Gazankulu, for example. Figure 10.1 shows the extent of privately owned areas dedicated to nature conservation and Kruger National Park, itself comprising 0.5 million ha.

Opportunity Costs of Redistributing State Land

From an economic efficiency perspective, land should be utilized to give maximum economic benefit, particularly where land is a scarce resource, with the role of the government being to attend to distributional issues of wealth and income distribution. In the Transvaal, the land available for alternative uses is appraised for use in small-scale agriculture or game ranching—each with its environmental, distributional, and income-generating potentials.

If land presently under nature conservation were redistributed to individual farmers, the likely returns to farming may be lower than for the present uses such as game ranching. Shackleton (1996) cites estimates that returns from commercial ranching are only 6% of those from game ranching on an area basis in the Central Lowveld region of the Transvaal. Further evidence of the potential income-generation capacity from nature conservation is available for the nearby Kruger National Park (Engelbrecht and Van der Walt, 1993). This study shows that revenue generated by Kruger National Park (in rands, R) was R90 million in 1993, with a gross margin of R40 million. Alternatively, arable farming could be carried out on only 25,000 ha of the park, the area with mean annual rainfall of greater than 600 mm.

The returns to the activities are shown in Table 10.1, illustrating that given present demand for the tourist resource and present agricultural yields, greater returns can be secured from maintaining the conservation area. This conclusion, that conservation is more "profitable" than agriculture, cannot be generalized and is crucially dependent on agricultural productivity, a factor which can be enhanced with infrastructural development. For Kenya, for example, an estimation of the potential economic returns to agriculture for the whole area of Kenya under national parks leads to the contrary conclusion, that given the demand for and potential returns on agricultural land, there is at least a financial case for conversion of at least part of that area (Norton-Griffiths and Southey, 1995).

In the case of Kruger National Park, Table 10.1 shows that the present R40 million return from economic activities in the park outweighs the estimated R18 to 22 million returns from arable agriculture and livestock. Moreover, maintenance of biodiversity confers unquantified benefits to society. In economic terms, there are significant option and existence values to conservation activities, values which have often been estimated to be at least as great as direct use or indirect use values, such as those highlighted in Table 10.1 (see Brookshire et

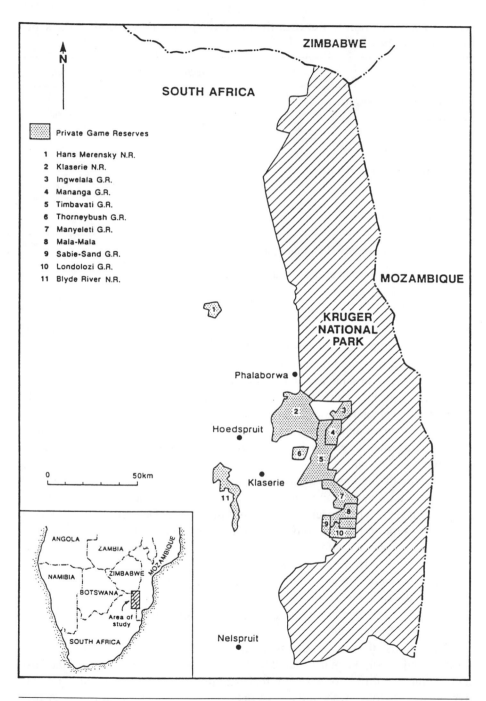

Figure 10.1 Nature conservation and land uses, Transvaal, South Africa.

Table 10.1   Economic Returns (in Rands [R]) from Alternative Uses of Kruger National Park

	Gross margin per ha × area	R million
Present conservation use		
Revenue		90
Less costs		50
Total annual return		40
Alternative use in agriculture		
Arable agriculture[a]	(R80–250) × 25,000 ha	2–6.25
Livestock	R8 × 2,000,000 ha	16
Total annual return		18–22.25

[a] Range based on improved and unimproved rain-fed agriculture.

Source: Based on Engelbrecht and Van der Walt (1993).

al., 1983, for example). However, the distribution of these existence and option values is also important to consider, in that the existence values accrue to non-users and their magnitudes are affected by human intervention and management of the habitats. The same additional non-quantified benefits of conservation, of option and existence value of wilderness for example, cannot equally be made by fenced private game reserves adjacent to Kruger National Park, as they have multiple objectives and management practices which diverge from conservation, by managing the ecosystem for game viewing or for breeding programs, for example.

Even for the national park area, where the estimated economic benefits of conservation are more than double the estimated opportunity costs, the analysis is based on two stylized scenarios that are unrealistic. The scenario for is based on the assumption that the conversion of the area to agriculture would destroy the habitat of the wildlife resources, and hence the associated non-marketed benefit to society from the existence of wildlife would be lost. Yet even if all land were converted to agriculture, it does not follow that the nature conservation interest would be lost completely: in Zimbabwe, wildlife resources in areas of extensive communally owned agricultural land have shown stable levels of population (Mbanefo and de Boerr, 1993).

Perceptions of communities in northern Transvaal are that private game reserve land should be redistributed to those households displaced over the past half century. A difficulty with any redistribution is the lack of recorded detail over exact previous ownership and of only traditionally enforced rights. A further difficulty is that ownership rights were lost incrementally over time. Thus, some parts of the population have been displaced more than once, hence losing a claim on a single specific area of land. Parts of the local communities in such areas as Transvaal have such grievances and are adamant that land rights are the key issue at the heart of their future development. Thus, although water is a critical resource in agricultural development in the semi-arid Transvaal, future assured supply is desired in the context of restored land rights. In these circumstances, the potential use of present private game reserves for agriculture is a relevant alternative, if not an immediate likelihood. Equity as well as efficiency are the driving forces behind future potential land reform as stated by the policy documents of the new government of national unity. The policy emphasizes the distribution of individual farm units.

Yet, as outlined above, communal land is not necessarily less efficiently or less sustainably managed than private land. In considering alternative uses of land, the additive benefits of sustainable use must also be considered. Thus, the estimates presented for Kruger National Park previously assume that agricultural and conservation activities are mutually exclusive. This is a principle inherent in national park systems in many parts of the world but one that does not apply to land not nationally designated as a site for conservation. Thus, integrating conservation and development objectives requires seeking opportunities for non-exclusive, sustainable utilization.

Alternative Property Rights and Sustainable Use of Private Land

The preceding analysis points to the need to consider alternative institutional arrangements for resolving the conflicts between promoting economic development on the one hand and interventionist wildlife management on the other. As indicated earlier, there is much experience, albeit controversial, with the role of communal management of wildlife resources in southern Africa. Yet even without consumptive use of wildlife resources through hunting revenues, opportunities exist for other multiple uses of such resources, for example, income generation through the non-consumptive utilization of wildlife (which makes up the greatest proportion of tourism) and the use of grass, wood, fruit, and other products either extracted or grazed *in situ*. These sources of local income can be significant. Further, economic activity associated with the tourism sector can be a source of income and employment.

Table 10.2 gives an estimate of the potential value of non-timber marketable products, derived from a study for the Transvaal region (Shackleton, 1996). The results show that the potential value of the marula fruit harvest, for example, constitutes a value of R5000 per hectare based on the prevailing market price of

Table 10.2   Estimated Potential Value of Secondary Products on a Sustained Yield Basis for Rangeland, Transvaal

Product	Price (R)	Yield/ha	Revenue/ha (R)
Fuelwood	0.08–0.16/kg	306 kg	24.8–50
Wild fruits	5/kg	1000 kg	5000
Thatch grass	0.5–2.0/bundle	1 bundle	0.5–2
Carving wood	30–70/tree	0.33 tree	10–23.1
Total (per ha)			5035–5075

Note: Figures are annual potential revenues based on sustainable yield. Annual value of carving wood per hectare is based on yield of 0.33 tree per hectare. Rotation of eight years required for this yield; therefore, the potential value only available in year 1, year 9, year 17, etc. The estimated potential fuelwood yield for private game reserve land is over four times that in communally owned land. Yield based on estimated standing biomass in conservation areas of 20.4 t/ha and mean annual increment of utilizable wood of 1.5%. The yield given for thatch grass is generally available only where rainfall is >600 mm.

Source: Based on Shackleton (1996).

the fruit, which is used as an ingredient in jam, fruit rolls, beer making, and other products. This value outweighs the other potential uses of such resources, including the sustainable yield of fuelwood, estimated as R24 to 50 per hectare. Results from research on secondary products from woodlands and rangelands in a range of low agricultural potential zones in Zimbabwe (Bojö, 1993) also show that these values are potentially significant. Fruit constituted one-third of the value of harvested products and wild foods almost 10%. Fuelwood in the research in Zimbabwe constituted only 20% of estimated total value.

The potential value approach is useful in showing orders of magnitude and the relative value of non-timber products. Research on the imputed value of non-marketed products was pioneered by studies such as that by Peters et al. (1989) in the Peruvian Amazon. Their study valued the utilized non-timber products in terms of the yield for each hectare less the extraction cost for an experimental site to demonstrate that the value of the products within a particular hectare of forest is greater than the timber value from clear-felling in that case. However, it has been observed that this benefit estimate is not transferable to other sites, as the location of secondary products fundamentally determines their value through access to markets. The criticisms of the Peters et al. study are also applicable to the estimates set out for the Transvaal. Unlike the Peruvian study, the estimates in Table 10.2 do not subtract extraction costs, which are not estimated in

Shackleton's study and are therefore not *gross margins* (revenue less total vari- able costs), the usual measure of the viability and profitability of any economic activity.

Other issues concerning the profitability of these activities include the im- pact of increased supply on market prices: the products generated from a single hectare would be unlikely to affect the price of fuelwood, carving timber, or marula. Opening up of reserve areas through significant redistribution of rights to utilize products would affect the relative prices of the resources in question, principally with falls in the wholesale price of the products. It is therefore not possible to extrapolate the estimated site-specific prices across large areas to derive aggregate potential revenues.

A further point is the sustainability and additivity of the different economic activities. In the case of carving wood, the yield estimate of one suitable tree in 3 ha would allow a sustainable harvest only every eight years. The *present value* estimate of sustained yield would show the relative value of the resource as carving wood to be less than shown in Table 10.2, as the yield would be avail- able only in sporadic years. A second issue regarding the potential value of the resource is the additivity of the individual elements: the impact of harvesting of any single element, particularly fuelwood, on the dynamics of the stand may mean that the different elements can be exclusive to an extent. An example is whether the fuelwood value includes harvesting of the species used for wood carving before they reach the necessary 6- to 8-m height.

In summary, there are various alternative uses of the bush resource, alterna- tives that would not require radical redefinition of the present property rights regime. The value of some elements of the potential resource has been quanti- fied for this region (Shackleton, 1996), but because any potential value is cru- cially dependent on extraction and marketing costs, the estimates of over R5000 per hectare should be viewed with caution. In comparison, the gross margins associated with private game reserve activities and commercial cattle ranching in the same region are estimated in the region of R1000 and R60 per hectare, respectively (Shackleton, 1996). Nevertheless, it has been shown by Bojö (1993), Shackleton (1996), and others that secondary, non-timber forest products can have significant potential value in woodlands and rangelands in southern Africa.

The benefits of nature conservation can be enhanced through tourism devel- opment, which promotes opportunities for employment and income generation in the formal sector. Employment opportunities are critical in the Transvaal, as opportunities for formal employment are extremely low. Among the population in the Gazankulu homeland, there is reported to be greater than 60% unemploy- ment in the formal sector. Overseas Development Institute (1994) cites estimates of only 53% of the "working" population employed in the formal sector for the whole of South Africa. Changes in employment practices can capture a greater proportion of tourist revenue locally, transfer skills, and hence stimulate economic development. Economic multipliers also tend to be high in such regions with low rates of leakage of income and possibly reduced rates of out-migration.

This section has demonstrated some of the values of the stock of products in areas such as semi-arid rangelands of northern Transvaal. This is only part of the range of benefits, including the use of medicinal plants and potential revenue from natural tourism (Cunningham, 1991; Koch, 1994). Other values of land are less tangible: the cultural and historical significance of land rights cannot be estimated in economic terms yet form the basis of a powerful policy objective in South Africa. Although this section has *demonstrated* the value of products existing in nature conservation land in the region, mechanisms for *appropriation* of these benefits require the strengthening of available markets for the products and rights to be established for their extraction. Such mechanisms can be based on traditional management of wildlife or extractive resources or can involve reinventing common resource management (see Gibson and Marks [1995], Anderson [1990], and Plotkin and Famolare [1992] for examples in other areas). Similarly with employment generation, the creation of employment opportunities requires institutional changes in labor markets and infrastructural support (see Gaiha [1993], for example). The fundamental issue of ownership of land is not addressed by such activities which illustrate the divergence between the narrow issue of land reform and redistribution and that of changing property rights and access to resources.

Conclusions

Policies to promote sustainable land use in South Africa can learn from the experience of other African nations. Many features of the South African situation, such as the economy, human resource base, and ecological diversity, make the task of implementing sustainable development less intractable than for many African countries at the equivalent stages of post-independence. However, this chapter has argued that sustainable development cannot proceed without addressing the central consideration of both the ownership of and the access to resources, instead of simply the efficient use of natural resources, and hence the maintenance of natural assets in their varied forms.

The argument for large-scale redistribution of land focuses on equity issues. On economic grounds, the transition from commercial to small-scale farming has not had detrimental effects on overall production and national food security in such other areas of Africa where similar transitions have been enacted (Lipton and Lipton, 1994). Homogeneous policies of land redistribution for agriculture which do not consider the agro-ecological diversity and potential of the land are not necessarily desirable (Williams, 1996). The case that the highest returns from arid and semi-arid zones come from agriculture is not proven. It is not, however, an argument against reform *of access* to such resources. As pointed out by Bromley (1994), land that changes ownership would not necessarily be farmed, and a market for land could be encouraged so that this land could return to high-value usage, with resulting compensation for displacement.

In the circumstances outlined in the Transvaal, alternative property rights could be instigated for the utilization of resources currently used for nature conservation. The potential value of some of these resources has been outlined in this chapter. The standard definition of "integrated conservation and development" is of sharing the benefits of utilization of the resource, whether these result in the consumptive or non-consumptive use of wildlife. The integration of community development with nature conservation requires multiple use of conservation areas with agriculture, extraction of resources, or schemes for capturing revenue from tourism. Equitable and sustainable use is essential if conservation and development are to be integrated.

Acknowledgments

The Centre for Social and Economic Research on the Global Environment is based at University of East Anglia and University College London and is a designated research center of the U.K. Economic and Social Research Council. The author thanks Katrina Brown, Tom Crowards, Dominic Moran, and Fraser Smith for comments on earlier drafts. This research was supported by the U.K. Overseas Development Administration. All views expressed in the chapter are those of the author alone.

References

Adams, J.S. and McShane, T.O. (1992) *The Myth of Wild Africa: Conservation without Illusion.* W.W. Norton, New York, NY.

Anderson, A.B., Ed. (1990) *Alternatives to Deforestation.* Columbia University Press, New York, NY.

Behnke, R.H. and Scoones, I. (1992) Rethinking Rangeland Ecology: Implications for Rangeland Management in Africa. Environment Department Working Paper 53. World Bank, Washington, D.C.

Bojö, J. (1993) Economic valuation of indigenous woodlands. In: *Living with Trees: Policies for Forestry Management in Zimbabwe.* P.N. Bradley, and K. McNamara, Eds. Technical Paper 210. World Bank, Washington, D.C., pp. 227–241.

Bromley, D.W. (1989) Property relations and economic development: the other land reform. *World Development,* 17: 867–877.

Bromley, D.W. (1994) The enclosure movement revisited: the South African commons. *Journal of Economic Issues,* 28: 357–365.

Bromley, D.W. (1995) South Africa: where land reform meets land restitution. *Land Use Policy,* 12: 99–103.

Brookshire, D., Eubanks, L., and Randall, A. (1983) Estimating option price and existence values for wildlife resources. *Land Economics,* 59: 1–15.

Carruthers, J. (1993) Police boys and poachers: Africans, wildlife protection and national parks, the Transvaal 1902 to 1950. *Koedoe: Research Journal for National Parks in the Republic of South Africa,* 36(2): 11–22.

Child, B. (1993) Zimbabwe's CAMPFIRE programme: using the high value of wildlife recreation to revolutionize natural resource management in communal areas. *Commonwealth Forestry Review,* 72: 284–296.

Cliffe, L. (1994) Zimbabwe's experience in promoting sustainable rural development and the implications for South Africa. In: *Sustainable Development for a Democratic South Africa.* K. Cole, Ed. Earthscan Publications, London, U.K., pp. 68–83.

Cunningham, A.B. (1991) Development of a conservation policy on commercially exploited medicinal plants: a case study from Southern Africa. In: *The Conservation of Medicinal Plants.* O. Akerle, V. Heywood, and H. Synge, Eds. Cambridge University Press, Cambridge, U.K., pp. 337–357.

Deininger, K. and Binswanger, H.P. (1995) Rent-seeking and the development of large scale agriculture in Kenya, South Africa and Zimbabwe. *Economic Development and Cultural Change,* 43: 493–522.

Ehrenfeld, D. (1992) The business of conservation. *Conservation Biology,* 6: 1–3.

Ellis, S. (1994) Of elephants and men: politics and nature conservation in South Africa. *Journal of Southern African Studies,* 20: 53–69.

Engelbrecht, W.G. and Van der Walt, P.T. (1993) Notes on the economic use of Kruger National Park. *Koedoe: Research Journal for National Parks in the Republic of South Africa,* 36(2): 113–120.

Favre, D. (1993) Debate within the CITES community: what direction for the future? *Natural Resources Journal,* 33: 875–918.

Gaiha, R. (1993) Design of Poverty Alleviation Strategies in Rural Areas. Economic and Social Development Paper 115. Food and Agricultural Organization, Rome, Italy.

Gibson, C.C. and Marks, S.A. (1995) Transforming rural hunters into conservationists: an assessment of community-based wildlife management programs in Africa. *World Development,* 23: 941–957.

Grohs, F. (1994) *Economics of Soil Degradation, Erosion and Conservation: A Case Study of Zimbabwe.* Vauk, Kiel, Germany.

Hackel, J.D. (1993) Rural change and nature conservation in Africa: a case study from Swaziland. *Human Ecology,* 22: 295–312.

Hardin, G. (1968) The tragedy of the commons. *Science,* 162: 1243–1248.

Hulme, M. and Kelly, P.M. (1993) Exploring the links between desertification and climate change. *Environment,* 35(6): 4–11, 39–45.

Kates, R.W. and Haarmann, V. (1992) Where the poor live: are the assumptions correct? *Environment,* 34(4): 4–11, 25–28.

Koch, E. (1994) Reality or Rhetoric? Ecotourism and Rural Reconstruction in South Africa. Discussion Paper 54. UNRISD, Geneva, Switzerland.

Lipton, M. and Lipton, M. (1994) Restructuring South African agriculture. *IDS Bulletin,* 25(1): 24–29.

MacKenzie, J.M. (1988) *The Empire of Nature: Hunting, Conservation and British Imperialism.* Manchester University Press, Manchester, U.K.

Mbanefo, S. and de Boerr, H. (1993) CAMPFIRE in Zimbabwe. In: *Indigenous Peoples and Protected Areas.* E. Kemf, Ed. Earthscan Publications, London, U.K., pp. 81–88.

Norton-Griffiths, M. and Southey, C. (1995) The opportunity costs of biodiversity conservation: a case study of Kenya. *Ecological Economics,* 12: 125–139.

Overseas Development Institute (1994) Economic Policies in the New South Africa. Briefing Paper 1994/2. ODI, London, U.K.

Pankhurst, D. (1995) Towards reconciliation of land issue in Namibia: identifying the possible, assessing the probable. *Development and Change,* 26: 551–585.

Peters, C.M., Gentry, A.H., and Mendelsohn, R.O. (1989) Valuation of an Amazonian rainforest. *Nature,* 339: 655–656.

Plotkin, M. and Famolare, L., Eds. (1992) *Sustainable Harvesting and Management of Rainforest Products.* Conservation International, Washington, D.C.

Shackleton, C.M. (1996) Potential stimulation of local rural economies by harvesting secondary products: a case study of the Central Transvaal Lowveld, South Africa. *Ambio,* 25: 33–38.

Stocking, M. and Perkin, S. (1992) Conservation with development: an application of the concept in the Usambara Mountains, Tanzania. *Trans. Inst. Br. Geogr.,* 17: 337–349.

Wells, M. and Brandon, K. (1992) *People and Parks: Linking Protected Area Management with Local Communities.* World Bank, Washington, D.C.

Western, D. and Wright, R.M., Eds. (1994) *Natural Connections: Perspectives in Community-Based Conservation.* Island Press, Washington, D.C.

Williams, G. (1996) Setting the agenda: a critique of the World Bank's rural restructuring program for South Africa. *Journal of Southern African Studies,* 22: 139–166.

Sustainable Development at the Village Community Level: A Third World Perspective

11

Victor M. Toledo

Centro de Ecología, Universidad Nacional Autónoma de México, Morelia, Michoacán, México

Abstract

World statistics indicate that almost half of the inhabitants of the planet are agricultural people. Of these, about 95% are inhabitants of Third World countries, the majority of which operate as households belonging to village communities. Examples of these community-based systems of resource management are particularly prevalent in rural areas of Mexico, India, and China and are found throughout the developing world.

This chapter is devoted to examining the new paradigm of sustainable development at the village community level. First, a political–ecological definition of sustainable community development is given. Then, nine ethno-ecological principles for sustainable community development are proposed and discussed. The chapter finishes by reviewing the aforementioned concepts in the light of grassroots experiences of sustainable development among peasant communities in rural Mexico.

Introduction

Rural or primary production, which includes agriculture, cattle raising, fisheries, and forestry, is the main human activity and the principal influence on our

1-57444-077-2/97/$0.00+$.50
© 1997 by CRC Press LLC

233

planet's ecology. Since 1987, when the World Commission on Environment and Development published the Bruntland report (WCED, 1987), the academic, scientific, and policy-making communities have focused considerable attention on the application of the concept of sustainability to the primary or agricultural sector. This interest took an international dimension through the governmental agreements derived from the Earth Summit held in Rio de Janeiro in 1992 (especially Agenda 21). However, despite this interest, most research has focused on agricultural systems and not on the social actors carrying out agricultural practices (see, for example, Faeth [1993], Ikerd [1993], and Farshad and Zinck [1993]). Only recently has a broader vision of sustainable rural development, encompassing social, cultural, economic, and political perspectives, begun to be elaborated (Altieri, 1992; Thrupp, 1993; SANE, 1994; ICSA, 1995). Under this broader perspective, village communities emerge as one of the central topics to be considered by researchers engaged in the new paradigm of sustainability (see the pioneering work of Agarwal and Narain [1989] in India).

This chapter concentrates on what are perceived as basic principles that should be addressed if sustainable development is to be achieved at the village community level. Village communities are the main social nuclei of the rural areas of the Third World, and they operate as social and productive units through which natural resources are appropriated. To complement and illustrate this, the chapter offers a brief review of the principal grass-roots experiences of sustainable development among indigenous and peasant communities of rural Mexico.

The Ecological Importance of Village Communities: A Global Perspective

World statistics indicate that almost half the inhabitants of the planet are agricultural people. In fact, 45% of the total human population has been recorded by the FAO as agricultural and 46% as the economically active population in agriculture (Table 11.1). Of these, about 95% are inhabitants of Third World countries and only 5% are in developed or industrialized nations.

In general terms, the rural or primary producers of the contemporary world belong to two main contrasting, and historically rooted, archetypes: peasants (pre-modern or traditional) and industrialized (or modern) (see Toledo [1995] for an ecologically oriented definition of these two prototypes). Although it is difficult to obtain with accuracy the proportion of the agricultural population corresponding to the two main types of rural producers, it can be estimated that between 60 and 80% are still peasants, almost peasant, or peasant-like producers. This estimation is based on the use of a key characteristic: small scale. The areas of the world where there is a clear small-scale predominance correspond

Table 11.1 **Agricultural Population and Economically Active Population in Agriculture (in Millions) in the Main Regions of the World**

	Population			Economically active population		
	Total	*Agricultural*	*%*	*Total*	*Agricultural*	*%*
World	5294	2389	45.1	2365	1101	46.6
Developed countries	1248	102	8.1	560	50	8.3
Third World countries	4045	2287	56.5	1765	1051	59.6
China	1139	786	67.4	680	458	67.5
India	853	535	62.7	322	214	66.5
Africa	524	346	66.0	243	154	63.2
Latin America	448	118	26.3	158	41	26.1

Source: *FAO Statistical Yearbook, 1991.*

principally to the Third World countries. In contrast, most of the industrialized nations of the North have countrysides dominated by medium- and large-scale agro-industrial units (with some notable exceptions, such as Portugal and southern Spain).

The statistical record shows that by 1990 around 1.2 billion rural people were involved in agricultural activities on areas of 5 ha or less (Table 11.2). This figure coincides with the last available world census of agriculture by the FAO in 1970, where more than 80% of all reported holdings were smaller than 5 ha (Wilken, 1987). Most of this peasant population develops its production activities not as socially isolated households but as familial nuclei belonging to specific village communities. They represent community-based systems of resource management, examples of which can be found in Egypt, Tanzania, Vietnam, Indonesia, Peru, Bolivia, Colombia, and especially Mexico, India, and China.

In Mexico, three million peasant households belong to nearly 30,000 ejidos (parcels of communal land) and indigenous communities (Table 11.3), whereas Indian peasantry manages natural resources distributed in over 500,000 villages (Table 11.4). In China, recent changes in the political economy have meant that rural areas have replaced village communities (and their households) as the main entities of productive life. In fact, during Mao's political period of rural collectivization, villages were substituted by brigades, communes, and cooperatives in order to take political control over peasantry. Under the new regime, the former system was abolished and its political and social functions transferred again to

Table 11.2   Peasant Small-Scale Population in 1990 (in Millions) for 17 Selected Countries

	Agricultural population in 1990 (millions)	% of holdings ≤5 ha	Small-scale agricultural population (millions)
Asia			
China	679.6	100	679.6
India	535.6	84	449.9
Pakistan	35.4	71	25.1
Iran	14.6	73	10.7
South Korea	9.6	100	9.6
Africa			
Algeria	5.9	80	4.7
Egypt	21.2	95	20.2
Tanzania	21.7	100	21.7
Latin America			
Brazil	36.5	44	16.1
Chile	1.7	49	0.8
Colombia	9.1	60	5.5
Costa Rica	0.7	43	0.3
Ecuador	3.2	71	2.3
El Salvador	1.9	87	1.7
Mexico	26.8	77	20.6
Peru	7.9	78	6.2
Venezuela	2.1	49	1.0
Total			1277

Note: Estimated from FAO statistics and other sources.

Table 11.3   Number of Holdings and Owned Areas in 1991 for the Peasant (Community Based), Private, and Mixed Sectors in Rural Mexico

	No. holdings	%	Area (million ha)	%
Community based	3,040,495	66.3	103.3	59.0
Private	1,410,742	30.8	71.7	40.9
Mixed	133,912	2.9	0.13	0.1
Total	4,585,149	100	175.1	100

Source: Seventh Agricultural, Forestry and Peasant National Census of Mexico.

Table 11.4   Distribution of Village Size in India

Village population	No. villages	Percent of total
Less than 200	120,073	21.6
200–499	150,722	27.0
500–999	135,928	24.4
1,000–1,999	94,486	17.0
2,000–4,999	46,892	8.4
5,000–9,999	7,202	1.3
Over 10,000	1,843	0.3
Total	557,137	100

Source: From Agarwal and Narain (1989).

the local government of the villages. By the end of 1983, the new system had been adopted by about 95% of over 120 million peasant households (Ling, 1991).

Defining Sustainable Development at the Village Community Level

Community development has been a main concern for academics for a long time, and recent contributions have emphasized two main social aspects: farmer participation and local empowerment (Farrington and Martin, 1988; Chambers et al., 1989; Thrupp, 1989). However, the new paradigm of sustainability has been adopted more as a technical or productive "factor" in order to generate an alternative or sustainable agriculture (see a critique in ICSA, 1995) than a holistically visualized perspective. For this reason, a broader and more integrative approach must focus on sustainable rural development and on the more specific topic of sustainable community development, which is the main concern in Third World countries.

It is possible to define sustainable community development as an endogenous mechanism that permits a community to take (or retake) control of the processes that affect it. This definition is derived from a general principle of political ecology which affirms that the fundamental reason why contemporary society and nature suffer generalized processes of exploitation and deterioration is the loss of control of human society over nature and itself. From this perspective, the history of humanity can be seen as a movement toward an ever greater loss of control over the processes that affect people and their surroundings, in contrast to the paradigm of "progress." In other words, self-determination and local empowerment, conceived as a "taking of control," have to be the central objectives in all community development. In fact, local self-reliance is visualized as the primary goal of sustainable development of rural communities, which have remained largely exploited by centralized urban powers throughout history (see a global review of this situation in Powelson and Stock, 1987).

It is possible to distinguish among six different types of process. The first action that a community should take is to establish control over its territory. This action implies the definition of the territory's boundaries and the legal or judicial recognition by national states of this territory as belonging to local communities and their members. The adequate or non-destructive use of natural resources (flora, fauna, land, water resources, etc.) that form part of a community's territory constitutes the second taking of control of a rural area. It succeeds through the design of, and a stake in, the execution of management plans for natural resources capable of standardizing and regulating agriculture, cattle raising, forestry, and fishing. Such a management plan would imply the elaboration of ecological diagnostic techniques, some kind of inventory, and, if possible, a geographical information system that would be able to evaluate the ecological resources of a community's territory.

Cultural control—where the community makes decisions that safeguard its cultural values—includes language, costumes, knowledge, beliefs, and lifestyles. The community must create mechanisms that guarantee cultural rescue and the consciousness of its members regarding the necessity of maintaining their main cultural traits (ethnic consistency). The improvement of the quality of life of the members of a community is a central task of all community development, and it agrees with establishing social control. This includes such aspects as nutrition, health, education, housing, sanitation, recreation, and information.

The regulation of economic exchanges that link a community with the rest of society or with international markets constitutes the taking of economic control. It implies confronting from a community perspective the external economic phenomena that affect a community's productive life, such as the politics of price fixing (for the market or the state), macroeconomic politics, subsidies, loans, etc. It implies attenuating the mechanisms that inhibit and castigate the productive sphere of a community.

Finally, the last dimension is the taking of political control. It implies a capacity for the community to organize itself socially and productively in such a way as to promulgate or ratify the norms, rules, and principles that govern a community's political life. The taking of political control would not succeed without the exercise of real community democracy. It implies decision making based on the consensus of the members of a community, the rights and aspirations of individuals and families, and the defense of the community for all.

Each of these six dimensions of community development (territorial, ecological, cultural, social, economic, and political) cannot easily exist without the others; put another way, the reclaiming of control must be integrated and complete, with these six dimensions included together. For example, it is not possible to maintain and defend culture while the destruction of natural resources persists, which in turn tends to affect the quality of life. However, the defense of culture and nature, the maintenance and improvement of the quality of life of the members of a community (the producers and their families), and the suppression of those economic injustices that perpetuate unequal exchange within society will prove difficult tasks if a true political organization does not exist. The taking of political control (that is, community democracy) is, without doubt, the pivotal action on which the other dimensions of control depend.

These six processes which form true sustainable community development can only succeed to the extent that the members of a community acquire, increase, and consolidate a community conscience. In the majority of Third World countries, rural communities find themselves permanently under siege by the destructive forces of "modernizing development," based on the destruction of nature and collective wealth, and the consecration of individual interests—forces that an industrial, technocratic, materialistic society increasingly imposes in all corners of the world (for the Indian case, see Agarwal and Narain [1989]). For this reason, initiatives of sustainable development must take into account the social situation of the communities to be considered. It is theoretically possible to find not only well-organized communities but also those clearly experiencing social disruption or degradation. In all these cases, however, the community itself must elaborate, as a prerequisite, a plan for its development. This is the essential instrument of a community's struggle, for its resistance to current disintegrative forces, and the departure point for integrating its actions.

Nine Ethno-Ecological Principles for Sustainable Community Development

The ethno-ecological perspective treats rural communities as cells within the social organism, entrusted to realize the appropriation of natural resources (represented by ecosystems) through agriculture, forestry, and fishing. It permits the derivation of a conjunct of new principles that sustain actions directed at the establishment of the six types of control mentioned above (Figure 11.1). This vision situates communities at the socio-ecological intersection, as entities pulled in two directions by the forces of nature and society. Such a tension is the result of the material exchanges that a community operates with nature (ecological exchanges) and society (economic exchanges), from which it derives its metabolism (Toledo, 1990). These principles are derived in turn from the special situation of communities, that is, from their specific eco-geographical position.

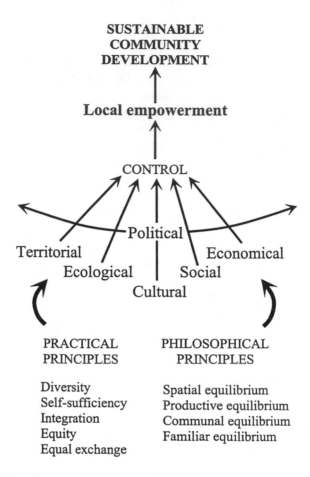

Figure 11.1   Principles and processes of sustainable development at the village community level as visualized in this chapter (for explanation, see text).

Of the nine principles described here, five form a group of practical principles, whereas the other four constitute philosophical principles. They are as follows:

1. *Diversity (biological, genetic, ecological, geographical, productive)*—This principle affirms an inherent trait in the peasant mode of appropriating nature; it lies in stark contrast to the homogenizing tendencies and specializations of development that prevail in a good part of the world today.

2. *Self-sufficiency (nourishment, energy, technology, economy, etc.)*—Advocated by numerous thinkers and philosophers (from St. Thomas to Gandhi), self-sufficiency implies the minimal dependence of a community on external sources. It must not be confused with autarchy, which implies total isolation.
3. *Integration (of productive practices, agricultural units, natural cycles, etc.)*—Presented in the ecological and economic rationale of the peasant, this principle tends to be overlooked and eroded by the dominant practices of modernization which seek community specialization and dependence.
4. *Equity (productive, of resources, of participation, etc.)*—In contrast with the numerous external mechanisms that promote the social and economic differentiation of communities (which give rise to different forms of social inequality), the maintenance of equity is an obligation of all community democracy.
5. *Economic justice*—Especially directed toward obtaining just prices for a community's commercial products, this principle seeks to confront the different mechanisms of unequal economic exchange which communities and their members have historically exploited for their markets. Economic justice aims to abolish unequal exchange between communities and their markets (economic exploitation), with the goal of avoiding the outbreak of unequal exchanges between communities and nature (ecological exploitation).

In contrast with the preceding five philosophical principles but also complementing them, the following four principles are organized around the idea of equilibrium:

6. *Spatial equilibrium*—This is fundamentally directed at attaining and guaranteeing what specialists call the "establishment of the land" through the harmonious management of different eco-geographical units that comprise a community's territory and their integration into productive processes. This principle, which is inherent in peasant rationality, also permits and promotes biological and genetic diversity and the equilibrium of material and energetic flows through ecosystems. It works through an equitable distribution of land dedicated to agriculture, ranching, and forestry in the communal territory. From another viewpoint, this principle is squarely opposed to all attempts to convert a community's natural resources into a monotonic "factory floor" for specialized production (the threat of the agro-industrial model). All this is a prerequisite for attaining the foregoing principles of diversity and self-sufficiency.

7. ***Productive equilibrium***—In the almost always problematic encounter that is established between use value and exchange value (i.e., between a productive rationality directed exclusively toward the subsistence of the producers and another that aims to overturn everything produced for the market), peasant wisdom has always searched for an equilibrium. This equilibrium aims, therefore, to adopt a strategy that maintains the value of exchange subsidiary to the interests and necessities of the community. Thus, it should avoid falling over the precipice of autarchy (total suppression of the value of exchange) and leaping into the inferno of the market economy (total suppression of use value). This principle appears, therefore, as an option between the false exit of a peasantry that "jumps back" from the precipice and a mercantilism that seeks daringly to destroy the self-regulatory capacity of communities, forming entities (both productive and consumptive) that are totally dependent on the market. The productive equilibrium seeks, therefore, to guarantee the reproduction of communities through a formula in which nature (ecological exchanges that guarantee self-sufficiency) operates like an ally that ensures safe navigation in the dangerous waters of the market.

8. ***Community equilibrium***—This principle seeks justice between the interests of the whole and the interests of its parts (i.e., between the collective or community interests and the interests of the families or individuals that form a community). It serves to avoid the excesses of collectivism and take advantage of the potential of individuals and their familial nuclei—a challenge that generally is evaded in development strategies.

9. ***Familial equilibrium***—Households being the main social and productive cells of any rural community, it is of enormous importance to guarantee a certain familial equilibrium. This principle, therefore, seeks to attain harmony between the individuals, sexes, and generations that integrate family units through the application of adequate standards of health, nutrition, hygiene, education, information, and recreation. Without the application of this principle to the "bricks of the building" (i.e., families within a community), every development strategy is condemned to collapse.

Implementing Sustainable Development at the Village Community Level: The Mexican Experience

In Mexico, two forms of community-based corporate ownership are currently recognized and supported by law: ejidos (*sic*) and comunidades (communities).

By 1991, 30,000 peasant communities (ejidos and indigenous villages) owned half of the Mexican territory (103 million ha), principally in forested areas (holding 70% of Mexico's forest land), agro-forested areas (coffee, vanilla, etc.), and agricultural lands, of which they hold 80%. Thus, while entrepreneurial farms dominate primary activities such as cattle ranching, irrigated areas, and some profitable branches of rain-fed agriculture, peasant communities are the main economic agents of forested regions and most rain-fed agricultural lands in Mexico.

In addition to the mestizos, or Spanish-speaking peasants, there are 54 different indigenous groups, speaking 240 languages other than Spanish, living in practically all the main natural habitats of the country, from tropical lowlands and temperate mountains to coastlands and desert or semi-desert areas.

Although the indigenous revolt in Chiapas in 1995 called world attention to Mexico's rural areas, a growing community-based movement in the Mexican countryside is searching for political objectives similar or close to sustainable development, as defined previously. The following is a brief review of the new ecologically inspired grass-roots social movements of rural communities (for further details, see Toledo, 1992).

Forestry Communities

According to the last national census, by 1991 over 15 million ha of tropical lowland and temperate mountain forests were under the dominion of peasant communities. Thus, peasant communities play a crucial role in the management of Mexican forests. Although community-owned forests were for centuries largely exploited by private and state companies, with virtually no benefits to the local owners, during the last two decades peasant and indigenous communities have been reappropriating their own forests. To date, there are several successful experiences of forestry management in many indigenous regions of the country, and an ambitious national training program in the sustainable management of community-owned forests was initiated in 1995 with the economic support of the World Bank.

Within this context, the experiences of the Maya of the tropical lowlands of Quintana Roo (Bray et al., 1993), of the Zapotec and Chinantec of the mountains of Oaxaca (Lopez-Arzola and Gerez, 1993), and of the Pur'hé of Nuevo San Juan in Michoacán (Alvarez-Icaza, 1993) are extremely important. They are model cases where economic and administrative efficiencies and ecologically sound forest management are successfully combined with community democracy, egalitarian ideology, and collective access to natural resources. Thus, they represent living examples of the idea of sustainable development as conceptualized previously. Although very little theoretical reflection on these experiences is presented here, it seems clear that they are consequences of the legal recognition of community-based property rights, plus certain positive changes in Mexico's forest concession policy, and a strength in society, culture, and ideology among the communities in question (Alcorn and Toledo, 1996).

In the national arena, the forest communities of Mexico also have been winning important political and organizational battles. For instance, the National Union of Forest Community Organizations, which is the largest community organization, with nearly 550 ejidos and comunidades covering a forest area of over 4 million ha, is demanding from the Mexican government changes in forest law, technical support, and economic incentives to realize sustainable management of the forests. Its strategy includes forestry management for timber and non-timber products, food self-sufficiency, and conservation of biodiversity, soils, water, and of the quality of life of local people.

Coffee-Growing Communities

Within the context of worldwide coffee production, Mexico, at present, stands fourth in terms of volume, fifth in terms of harvested surface, and ninth in terms of yield performance. It is estimated that the number of coffee producers has reached approximately 200,000. This accounts for a population of 1.5 million economically involved in the cultivation of coffee.

In Mexico, 70% of coffee production is performed by social groups pertaining to rural communities (ejidos and indigenous communities), while the remaining 30% is carried out by private producers. A large extent of this community-based sector is made up of indigenous producers. In fact, the cultivated area spans over nearly 700,000 ha, 90% of which corresponds to small holdings covering less than 5 ha, while 70% pertains to holdings extending over less than 2 ha. Thus, in Mexico, traditional growers dominate in terms of population and of planted land in coffee. They maintain multilayered, shaded coffee agro-forests, which contrast with the modern, agro-industrial, unshaded coffee plantations, which utilize chemical inputs and year-round labor.

As opposed to countries such as Brazil, which is the world's largest coffee producer and where production systems are made up of large-size private farms located in flat lowlands under shadeless monoculture (or direct sunlight), using high doses of agro-chemical nutrients, coffee in Mexico is basically produced on the slopes of mountain ranges spanning the southern and central parts of the country. Plots sown with coffee are found under a canopy of trees and worked by small (and in some cases extremely small) groups of producers generally from indigenous or peasant communities (Moguel and Toledo, 1995). This situation springs from the country's agrarian and cultural history, where native wisdom has literally appropriated an exotic crop adapted and suited to the native agro-forestal systems. Some experts hold that coffee, on its arrival from Africa via Europe at the end of the 18th century, substituted for cocoa within the complex Meso-American agro-forestal systems—a hypothesis yet to be proven.

As a result of these patterns, small coffee producers in Mexico (and especially those with an indigenous background) have never let coffee grow alone: they have always planted it alongside numerous other plant species (mostly useful in their own right, commercially or for subsistence) within what is technically called multiple-crop cultivation. An interesting consequence is that Mexico is the world's primary exporting country for organic coffee (accounting for one-fifth of the total volume) and that a substantial part of this production, carried out without using agro-chemical additives, is performed in the indigenous communities of Oaxaca, Chiapas, and other states. More interesting still is that the National Coordinator of Coffee Organizations (CNOC), which is the most important organization of small coffee growers in Mexico, is adopting a development policy based on the new paradigm of sustainability. Through this policy, CNOC is promoting and/or reinforcing ecologically and socially sound coffee-growing systems among its roughly 75,000 affiliated producers.

Fishery Communities

Mexican fishery production is carried out by fishing cooperatives using both industrial/modern, and artisanal/traditional methods. Similar in philosophy and often identical in organizational structure to the agricultural ejidos, fishing cooperatives represent the main productive organizations in charge of the management of oceans, coastal lagoons, and fresh waters. While industrial exploitation dominates in the northwest, around Baja California, principally artisanal, small-scale fisheries extend around most of the Gulf of Mexico, Caribbean, and southern Pacific coast regions.

During the last two decades, Mexican coastal ecosystems have suffered increasing coastal pollution and accelerated destruction of estuarine and coastal habitats. These processes have been especially strong in the tropical waters of the south, where oil extraction, urban coastal development, and tourism megaprojects have contributed to the depletion of marine and coastal resources.

As a response, during the last few years, an important social movement of traditional or artisanal fishing villages has initiated a variety of political actions against oil interests and other industrial polluters in defense of its natural resources. Through the National Movement of Riverine Fishermen, these fishery communities have fought against oil pollution in different regions of Michoacán, Veracruz, Tabasco, and Campeche. In addition, while negotiating with the

Mexican government, they have initiated new development programs inspired by the principles of sustainability.

Conservation-Oriented Communities

In the country with the third richest biological diversity on earth (Mittermeir, 1988; Rammamorthy et al., 1993), the geographical distribution of peasant communities coincides with most of the areas of extraordinarily high biodiversity. Nearly 80% of the territory of the protected areas established by the Mexican government is inhabited by indigenous people, and half of the 30,000 ejidos and indigenous communities are distributed in the ten most biologically rich states of Mexico. Thus, peasant community participation in the design and management of Mexican natural areas is a basic premise.

In addition, it is of great importance to stress that the main Biosphere Reserves of Mexico are surrounded by indigenous communities and that most of these communities claim active participation in the management of protected areas. This situation is especially notable in the south (Montes Azules, Calakmul, Santa Marta) but also exists in the central (Manantlán) and northern (El Pinacate) portions of the country (see Figure 11.2).

In this context, it is important to point out the processes initiated by indigenous communities, environmental organizations, conservationist groups, and Mexican scientists to create community-managed ecological reserves through the implementation of sustainable development programs. This is the case in Los Chimalapas (a region of southeastern Mexico in the heart of the Isthmus of Tehuantepec), in the reserve of the monarch butterfly in the state of Michoacán, and in the Calakmul Biosphere Reserve in Campeche (Figure 11.2). In Calakmul, for instance, a project financed by the Global Environment Facility of the World Bank, the governments of Mexico and Canada, and several academic institutions is being carried out in conjunction with the regional indigenous organizations around the reserve. This development project is adopting a multi-use strategy through which villages manage tropical forests for timber and non-timber products (chicle, allspice, palm leaves, and wild fauna), maize fields, agro-forestry, home gardens, intensive cattle areas, apiculture, and ecological and archaeological tourism (Boege, 1995).

Concluding Remarks

Peasants and their communities have historically been an obstacle to the implantation of the industrial or Western civilizatory model in both its capitalist and socialist versions. In fact, the destruction of peasantry has been a central objec-

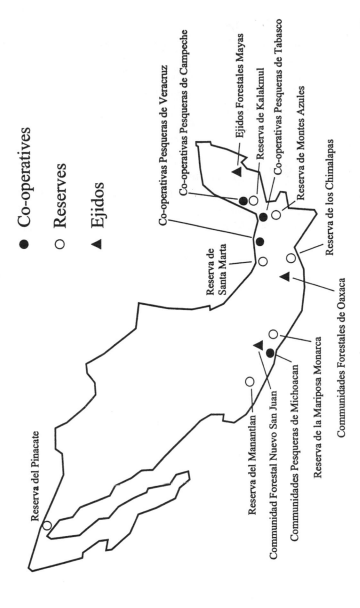

Co-operatives
Reserves
Ejidos

Reserva del Pinacate

Reserva del Manantlan
Communidad Forestal Nuevo San Juan
Communidades Pesqueras de Michoacan
Reserva de la Mariposa Monarca
Communidades Forestales de Oaxaca

Reserva de Santa Marta

Co-operativas Pesqueras de Veracruz
Co-operativas Pesqueras de Campeche
Ejidos Forestales Mayas
Reserva de Kalakmul
Co-operativas Pesqueras de Tabasco
Reserva de Montes Azules
Reserva de los Chimalapas

Figure 11.2 Map of Mexico showing the location of reserves, cooperatives, and ejidos.

tive in the modernization processes of rural life in Europe, Japan, North America, and the former Soviet Union. For this reason, rural modernization has been a compulsive process of the substitution of peasants' small-scale, solar-energized productive units for entrepreneurial or collective, large-scale, fossil-dependent holdings. The intensity of this phenomenon worldwide can be certified by the statistics regarding the number of people engaged in agricultural activities in each country.

As large-scale agriculture has been introduced in peasant-dominated rural areas, the number of producers has progressively declined. As a consequence, while 1990 agricultural populations were 13% of the total in the former Soviet Union, 9% in Europe, 3% in the United States, and 2% in Canada, this percentage was 68% in China, 63% in Africa, 53% in Asia, and about 30% in Latin America (FAO, 1991).

In contrast, a new strategy of sustainable development at the community level can be converted into appropriate options for rural modernization based on the evolution—not the substitution—of peasant practices and the adequate management of natural resources. Ultimately, the ethno-ecological approach demonstrates the existence of a certain ecological rationality of the peasant (Toledo, 1990). The ideas presented in this chapter, united with the experiences that are being developed in various parts of the Third World (and especially in India and Mexico), suggest that this is not only desirable but also feasible.

References

Agarwal, A. and Narain, S. (1989) *Toward Green Villages: A Strategy for Environmentally Sound and Participatory Rural Development.* Center for Science and Environment, New Delhi, 52 pp.

Alcorn, J.B. and Toledo, V.M. (1996) Resilient resource management in Mexico's forest ecosystems: the contribution of property rights. In: *Linking Social and Ecological Systems for Resilience and Sustainability.* F. Berkes and C. Folke, Eds. Beijer International Institute for Ecological Economics, Stockholm, Sweden, in press.

Altieri, M.A. (1992) Sustainable agriculture development in Latin America: exploring the possibilities. *Agriculture, Ecosystems and Environment*, 39: 1–21.

Alvarez-Icaza, P. (1993) Forestry as a social enterprise. *Cultural Survival Quarterly*, 17(1): 45–46.

Boege, E. (1995) La Reserva de la Biosfera de Calakmul, México (The Calakmul Biosphere Reserve). Unpublished manuscript, 40 pp.

Bray, D.B., Arreón, M., Merino, L., and Santos, V. (1993) On the road to sustainable forestry. *Cultural Survival Quarterly*, 17(1): 38–41.

Chambers, R.G., Pacey, R.A., and Thrupp, L.A., Eds., (1989) *Farmer First: Farmer Innovation and Agricultural Research.* Intermediate Technology Publications, London, U.K., 219 pp.

Faeth, P. (1993) An economic framework for evaluating agricultural policy and the sustainability of production systems. *Agriculture, Ecosystems and Environment,* 46: 161–173.

FAO (1991) *Statistical Yearbook, 1991.* Food and Agricultural Organization of the United Nations, New York, NY.

Farrington, J. and Martin, A. (1988) Farmer Participation in Agricultural Research: A Review of Concepts and Practices. Agricultural Administration Occasional Paper No. 9. Overseas Development Institute, London, U.K.

Farshad, A. and Zinck, J.A. (1993) Seeking agricultural sustainability. *Agriculture, Ecosystems and Environment,* 47: 1–12.

ICSA (1995) *Seeds of the Future: Sustainability and Natural Resources in the Americas.* Interamerican Council for Sustainable Agriculture, Mexico City, Mexico, 72 pp.

Ikerd, J.E. (1993) The need for a system approach to sustainable agriculture. *Agriculture, Ecosystems and Environment,* 6: 147–160.

Ling, Z. (1991) *Rural Reform and Peasant Income in China.* St. Martin Press, New York, NY, 191 pp.

Lopez-Arzola, R. and Gerez, P. (1993) The permanent tension: the forest communities of Oaxaca. *Cultural Survival Quarterly,* 17(1): 42–44.

Mittermeir, R.A. (1988) Case studies from Brazil and Madagascar and the importance of megadiversity countries. In: *Biodiversity.* E.O. Wilson, Ed. National Academy Press, Washington, D.C. pp. 145–154.

Moguel, P. and Toledo, V.M. (1995) El café en México: sustentabilidad y resistencia campesina e indígena (Coffee in Mexico: sustainability and peasant and indigenous resistance). *La Jornada del Campo,* 40: 6–8.

Powelson, J.P. and Stock, R. (1987) *The Peasant Betrayed: Agriculture and Land Reform in the Third World.* Lincoln Institute of Land Policy, Cambridge, MA, 302 pp.

Rammamorthy, T.P. et al., Eds. (1993) *Biological Diversity of Mexico: Origins and Distributions.* Oxford University Press, Oxford, U.K., 812 pp.

SANE (1994) *Creating Agroecological Lighthouses Around the World.* Sustainable Agriculture Networking and Extension, Berkeley, CA, 10 pp.

Thrupp, L.A. (1989) Legitimizing local knowledge: from displacement to empowerment for Third World people. *Agriculture and Human Values,* Summer Issue: 13–24.

Thrupp, L.A. (1993) Political ecology of sustainable rural development: dynamics of social and natural resource degradation. In: *Food for the Future: Conditions and Contradictions of Sustainability.* P. Allen, Ed. John Wiley and Sons, New York, NY, pp. 47–73.

Toledo, V.M. (1990) The ecological rationality of peasant production. In: *Agroecology and Small-Farm Development.* M. Altieri and S. Hecht, Eds., CRC Press, Boca Raton, FL, pp. 53–60.

Toledo, V.M. (1992) Toda la utopía: el nuevo movimiento ecológico de los indígenas y campesinos de México (All Utopia: the new ecological movement of indigenous people and peasants in Mexico). In: *Autonomía y Nuevos Sujetos Sociales en el Desarrollo Rural (Autonomy and New Social Subjects in Rural Development).* J. Moguel et al., Eds. Siglo XXI Editores, Mexico City, Mexico, pp. 33–54.

Toledo, V.M. (1995) Peasantry, Agroindustriality and Sustainability: The Historical and

Ecological Grounds of Rural Development. Working Paper 3. Interamerican Council on Sustainable Agriculture, Mexico City, Mexico, pp. 1–27.

WCED (1987) *Our Common Future.* World Commission on Environment and Development and Oxford University Press, Oxford, U.K., 400 pp.

Wilken, G.C. (1987) *Good Farmers: Traditional Agricultural Resource Management in Mexico and Central America.* University of California Press, Berkeley, CA, 302 pp.

Energy, Environment, and Development: Egypt, South Africa, and China

12

Kathleen L. Abdalla*
United Nations Economic and Social Commission for Western Asia,
Amman, Jordan

Abstract

This chapter considers the problem of encouraging economic growth, which is usually associated with greater energy use, and at the same time preventing environmental degradation. This dilemma is basic to all aspects of sustainable development policy but is perhaps most crucial for those associated with energy, given the strong link between economic growth and energy use, especially in developing countries.

Promoting sustainable development requires a combination of policies aimed at energy demand and supply, including broad-based policies such as pricing, institutional reform, social policies, and those aimed at specific energy-consuming sectors. Energy demand and supply patterns and the conditions of developing countries are examined here with such policies in mind. Three countries with different economic needs, energy requirements, and environmental problems, namely Egypt, South Africa, and China, illustrate the broad range of different policies required for sustainable development in developing countries.

---

* The views expressed here are those of the author and do not necessarily reflect those of the United Nations.

1-57444-077-2/97/$0.00+$.50
© 1997 by CRC Press LLC

Introduction

A major problem facing developing countries today is that while energy is necessary for development, it is a major source of environmental degradation. Economic growth necessitates the use of energy, while policies aimed at safeguarding the environment call for limitations on energy use. Expanding the productive base of an economy often implies an increase in energy use, and many developing countries cannot afford the expensive technology required for environmental restoration or efficient use of energy. The environment is an example of a public good that is overused due to an inability to regulate its use. While developed countries are attempting to introduce ways and means to limit and rationalize environmental use in an efficient way, for most developing countries such efforts are secondary in importance to meeting the basic needs of the population. In terms of priority, economic development comes first, often at the expense of the environment.

Complicating the issue further is the widely held view in the developing world that environmental degradation (especially the thinning of the ozone layer and the climatic warming trend) is a problem precipitated by the developed world. Certainly with regard to carbon energy use, the developed countries consume the bulk of fossil fuel and are thus the major contributors to global environmental degradation. In 1994, fossil-fuel consumption for the developed countries totaled nearly 5 billion tons of oil equivalent (TOE) compared with 2280 million TOE consumed by the developing countries (British Petroleum, 1989–95).

Per-capita energy consumption of primary, commercially traded energy in North America during 1994 was greater than 6 TOE, slightly above 3 TOE in Europe, and a little above 0.5 TOE in the rest of the world. Indeed, the importance of environment and energy policies in developing countries lies in the potential adverse effects of energy use on the environment. Growth in worldwide energy demand will occur as developing countries prosper and grow in the coming century and as their populations increase. Notorious pollution problems in cities such as Cairo, Bangkok, and Mexico City foretell the future if appropriate measures are not taken.

This chapter reviews world energy demand and supply prospects, as well as those in developing countries, with a view to the adoption of appropriate policies given such climates. Energy policies aimed at sustainable development goals can be directed toward influencing demand patterns or the supply of energy of a given country. For example, the relative control over emissions in developed countries is due in large part to increased energy efficiency. Demand-oriented policies can affect efficiency, and this holds great potential for developing countries. On the supply side, each fuel type is associated with different levels of environmental degradation, but the cleanest burning fuels may have other disadvantages for a given country. Ideally, an overall strategy will include both types of policies.

Three "developing" countries are considered here. They are characterized by different economic, political, and institutional climates, as well as by different levels of energy demand and supply situations. Egypt, South Africa, and China are examined with a view to assessing their current and potential success in promoting sustainable development in their respective energy policies. These countries provide interesting and illustrative examples of energy demand and supply patterns in developing countries which have been affected by various economic, political, and institutional factors. Egypt is faced with a growing population, low per-capita income, limited agricultural land, and a severe local environmental problem, but the country possesses oil and natural gas as well as a large hydroelectric facility. South Africa's energy situation has been influenced by its past political system, and its present-day energy policies must deal with localized pollution problems associated with the widespread use of coal in the majority of households. Although a minority of people in South Africa enjoy an "industrialized" standard of living, most people do not. South Africa is, therefore, in some ways a developed country and in others a developing country; therefore, it faces environmental problems similar to those in developed countries, but it faces many economic development problems found in developing countries. China's population and current economic growth imply a potentially significant contribution to global environment problems. Its growing use of energy and the importance of coal to its economy draw attention to its energy policies from the rest of the world. While each of the three countries considered here is different in terms of population, lifestyles, path of development, and resource endowments, the basic dilemma between economic development and a clean environment is common to all.

Sustainable development requires that economic growth occur with minimal degradation to the environment and that the costs of such degradation be accounted for in the production of goods and services. Developing countries are in a unique position to avoid some of the environmental pitfalls of development experienced by developed countries during industrialization. Trade-offs often exist between rapid economic growth and environmental degradation. Sustainable development may require a slower rate of economic growth since efforts and techniques associated with clean energy use are typically more expensive. In addition, the basic dilemma of sustainable development—that energy is necessary for economic growth but its use contributes to global and local pollution—is made more difficult in developing countries by impediments caused by institutional and political factors.

Energy Demand

The world demand for energy is expected to grow well into the next century in both developed and developing countries. Table 12.1 shows recent demand forecasts to the year 2010. Energy demand is expected to increase by an average annual rate of 3.1% during the first decade of the 21st century, which, although not as high as growth in the 1970s, represents an increase over the world growth in energy demand during the 1980s and that forecasted for the 1990s. Although energy demand and economic growth have been positively related historically, the developed countries of the Organization for Economic Co-operation and Development (OECD) have been successful in delinking these factors through a combination of demand management policies and the adoption of energy-efficient technologies. As a result, energy intensities (the ratio of energy use to gross national product) of developed countries have declined in recent years. Growth in energy demand in developed countries has steadily decreased since the 1970s and is expected to increase only slightly in the first decade of the 21st century. Developed countries' share of world consumption is expected to fall from 55% in 1992 to 51% in 2010. Growth in energy demand in Eastern Europe and the former Soviet Union has closely mirrored changes in economic conditions. Energy use will probably increase significantly in the coming years as the economies of this group of countries recover from a recent recession.

The largest increases in energy demand are expected to occur in the developing countries as economic growth occurs and populations increase. Although not expected to be as high as in the 1970s, growth in all developing areas is

Table 12.1   Actual and Projected Energy Demand Growth Rates (Average Annual Change in Percent)

	1970–1980	*1980–1990*	*1990–2000*	*2000–2010*
OECD[a]	3.3	2.8	2.3	2.4
Eastern Europe and former Soviet Union	3.3	1.6	–2.7	4.1
Non-OECD Asia	6.0	7.2	6.8	5.5
Middle East	5.2	0.9	4.0	3.7
Africa[b]	4.2	1.4	2.6	2.9
Central and South America	5.6	1.1	3.7	3.6
Total	3.6	2.7	2.3	3.1

[a] Including Turkey.
[b] Including Egypt and North Africa.

Table 12.2 Actual and Forecasted Regional Shares of World Energy Supply (percentages[a]) Calculated from Data in USEIA (1995)

	1992	2010
OECD[b]	54.7	50.7
Eastern Europe and former Soviet Union	18.2	15.8
Non-OECD Asia	16.3	22.1
Middle East	3.5	3.9
Africa[c]	3.1	3.1
Central and South America	4.1	4.4

[a] Columns may not add up to 100 due to rounding.
[b] Including Turkey.
[c] Including Egypt and North Africa.

forecast to increase significantly and in all cases at higher rates than the OECD countries. The largest increase was projected to occur in the non-OECD countries of Asia as rapid economic expansion continues there. China, for example, is expected to become a major energy user. Energy use for the 1990s is estimated to be growing at a rate of 6.8% per year and is forecasted at 5.5% per year between the years 2000 and 2010. Although lower than growth during the 1980s, this still represents a significant increase, especially given Asia's relative economic importance in the world. Asia is also expected to consume a greater share of the world's energy in 2010 compared with its share in 1992, as shown in Table 12.2.

The other developing regions, in which the growth of energy demand averaged around 5% per year in the 1970s, witnessed a severe drop in demand growth in the 1980s, which has rebounded in the 1990s and is forecast to average greater than 3% during the first decade of the 21st century. The Middle East, Africa, and Central and South America together accounted for 11% of the world's total energy consumption in 1992, and their share is projected to increase only slightly by 2010. However, this will still represent a significant increase in absolute terms over 1992 levels.

Many factors will contribute to the growth in energy demand in developing countries. Structural adjustment policies, encouragement of the private sector, and the participation of developing countries in the General Agreement on Tariffs and Trade and the World Trade Organization are expected to contribute to economic growth and thus energy demand in developing countries. Along with population growth in developing countries, which is forecast to grow from 4.5 billion people in 1995 to 5.8 billion in 2010 (United Nations, 1995), increased energy use is associated with economic development as lifestyles change and more energy-intensive products, often luxury items, are consumed in greater quantities. Real gross domestic product (GDP) growth for the developing countries as a group is projected at an annual average rate of 4.9% from 1995 to 2004, while per-capita GDP is projected to increase by 3.3% per year during the same period (World Bank, 1995a). Developing countries have been characterized by high energy income elasticities (i.e., as incomes increase, energy use increases by a greater proportion). In 1990, the energy income elasticity was 1.2 for developing countries compared with 0.5 for developed countries, which means that the use of energy will increase by 119% for every 100% increase in GDP, although both figures are expected to decrease by 2010 (USEIA, 1995).

The migration of large numbers of people from rural areas to urban areas usually precipitates increased energy consumption, and this trend is significant in developing countries. The urban population of the developing countries as a whole is expected to increase from 1.9 billion people in 1995 to 3.1 billion in 2010, or by an average annual rate of 3.3% (United Nations, 1995). The percentage of the developing countries' population living in urban areas is expected to increase from 41 to 52% during the same period. A switch from non-commercial energy to commercial energy occurs as the population moves to urban areas and per-capita energy consumption increases (Jones, 1991). For example, urban dwellers rely on transportation networks in traveling from the workplace to the home every day, while the rural population tends to live closer to the workplace. Goods and services must be transported to city markets to meet the needs of an urban population, while in rural areas many of the goods consumed are made in the home. In addition, an urban population usually implies an industrialized economy in which more energy is used by a sophisticated industrial sector.

This increased energy use, although necessary for economic growth, also means that carbon emissions and other pollutants will be greater as well, with damaging effects on the environment at the local, regional, and global levels. Given projected energy consumption increases, carbon emissions are expected to increase worldwide from about 6 billion metric tons in 1992 to 8 billion in 2010 (USEIA, 1995). Although the exact ramifications of these emissions at the global level are a matter of controversy, the effects at the local and regional levels are clear to most urban dwellers. For development to be sustainable, the environment must be considered an asset rather than a free good to be consumed without restraint. The challenge faced by developing countries today is to undertake ways and means of promoting economic growth while at the same time ensuring adequate protection of the environment. In the case of energy consumption, its use is necessary for growth but usually causes environmental degradation at the same time.

While an array of policy measures are available for the control of energy demand and promoting the efficient use of energy, many such measures may not be effective when applied in developing countries for a variety of reasons (Abdalla, 1994). For example, measures undertaken in developing countries to ensure that the consumer bears the cost of environmental degradation often require a degree of institutional support not available in most developing countries. In addition, the complexity of such arrangements may make them impractical, as might the specific political impediments which vary from country to country. A seemingly straightforward law, for instance, prohibiting or limiting certain emissions may be too costly to enforce in developing countries. Creating a "market" in which pollution rights are exchanged, another solution widely promulgated in developed countries, may face problems in countries where property rights are not widely recognized.

Perhaps the most effective measure that a developing country can undertake to constrain the consumption of energy is with regard to subsidies and the price charged to consumers. Many governments are recognizing that the subsidization of energy and energy products leads to their overuse, which in itself is not an efficient way to allocate resources, and causes artificially high emissions. With the assistance of various international institutions, these subsidies are gradually being reduced in many developing countries, along with subsidies on other consumer goods. Pricing energy according to its true costs, including the cost of environmental damage, effectively controls its use without a costly enforcement control apparatus, without additional institutional arrangements, and without introducing unfamiliar mechanisms that may be resisted by the public, depending on cultural mores and values. In addition, the pricing mechanism affects consumption by all consuming sectors rather than focusing only on one type of consumption, as do many rules and regulations (e.g., mandating the use of catalytic converters in automobiles).

Additional options include regulations aimed at controlling energy consumption in specific consuming sectors. Energy consumption in the transportation sector, for example, is affected by measures undertaken by governments to encourage mass transportation, specific regulations regarding automobile emissions, and regulations regarding the lead content of gasoline. Industrial sector consumption can be affected by a variety of rules and regulations to control emissions. Household consumption of energy can be influenced by mandating energy-efficient appliances and by consumer awareness campaigns specifically dealing with energy issues.

A more efficient use of energy means that less energy will be used at a given level of production or a greater production level will be feasible with a given level of energy use. Reduction or elimination of waste in all sectors can achieve substantial savings in energy in developing countries. In the production of electricity, for example, it is estimated that the developing countries are 18 to 44%

less efficient than the developed countries and that transmission and distribution losses compound the problem (Levine and Meyers, 1992). Other estimates point to the potential of increasing the standard of living in developing countries toward those of the developed countries with only a 30% increase in per-capita energy use, if OECD standards of energy efficiency are achieved (Goldemberg et al., 1986). This would assist developing countries in moving toward a high energy-efficient standard of living without incurring the significant increases in energy use that have characterized industrialization during the past 200 years. Energy efficiency requires changes in consumption patterns in every consuming sector. The household sector can benefit from adopting such changes as the use of more efficient appliances, the use of insulation in cold weather, and solar-powered devices where feasible. Energy efficiency in the transportation sector is enhanced with a well-planned mass transportation system and the use of fuel-efficient vehicles, while the industrial sector would benefit from new technologies dependent upon the production process and type of goods produced. Efficiency in the production of electricity has a widespread influence on overall energy efficiency since electricity is widely utilized in households, industries, commercial establishments, and government.

The fuel mix, or combination of fuels, used by a country can also contribute to sustainability. Policymakers in developing countries can target certain energy users or certain sources of energy through the types of policies undertaken. The financing of mass transit systems is one example of a selective policy, and the regulation of emissions allowed from automobiles and other vehicles is another. Direct legislation mandating limits on industrial pollution may be adopted as well. Giving tax incentives to households for energy-efficient activities has proved successful in developed countries, although it may be difficult to implement in developing countries. Other policies, such as the adoption of an extensive mass transport system, may be too expensive to finance. Energy-efficient technologies, especially those relevant to the industrial sector, may require a transfer of technology from developed countries not forthcoming due to institutional impediments, often including the lack of well-defined property rights. Thus, while opportunities exist for a more efficient use of energy, governments in developing countries may find it difficult to encourage energy consumers in all sectors to change their practices and habits.

Other effective measures recommended for developing countries but geared more toward long-term demand include programs to encourage rural development, population control, the general education of women (who are the main energy product decision makers in households), as well as the education of the population in general regarding efficient energy uses. These measures, often undertaken for other general sustainable development goals, also have the potential to curtail energy consumption in the long run.

Energy Supply

The supply of energy in a given developing country often reflects traditional energy demand patterns and the types of energy available within its boundaries. Total energy supplied in developing countries usually includes a significant amount of non-commercial energy (i.e., energy not traded), such as firewood gathered by members of the household. As populations move to urban areas, non-commercial energy use decreases. Table 12.3 shows the types of fuel used worldwide and by developed and developing countries in 1990 and forecast to 2010. While total energy use is projected to grow over the 20-year period, fuel

Table 12.3 Actual and Forecasted[a] Energy Supply in Million Tons of Oil Equivalent, 1990–2010 (Percentages in Parentheses)

	1990		2000		2010	
World						
Oil	3,302.4	(39.2)	3,817.1	(38.9)	4,422.0	(38.4)
Natural gas	1,756.1	(20.8)	2,175.6	(22.2)	2,600.0	(22.6)
Coal	2,241.5	(26.6)	2,451.2	(25.0)	2,878.0	(25.0)
Nuclear	495.1	(5.9)	578.0	(5.9)	595.1	(5.2)
Renewables	639.0	(7.6)	685.4	(7.0)	1,002.4	(8.7)
Total	8,429.3	(100.0)	9,819.5	(100.0)	11,504.9	(100.0)
Developed countries						
Oil	2,456.1	(39.0)	2,497.6	(36.8)	2,863.4	(37.4)
Natural gas	1,502.4	(23.9)	1,822.0	(26.9)	2,126.8	(27.8)
Coal	1,419.6	(22.5)	1,363.4	(20.1)	1,439.0	(18.8)
Nuclear	468.3	(7.4)	541.5	(8.0)	536.6	(7.0)
Renewables	456.1	(7.2)	558.5	(8.2)	678.0	(8.9)
Total	6,295.1	(100.0)	6,782.9	(100.0)	7,651.2	(100.0)
Developing countries						
Oil	846.3	(39.7)	1,319.5	(43.5)	1,558.5	(40.4)
Natural gas	253.7	(11.9)	353.7	(11.6)	473.2	(12.3)
Coal	822.0	(38.5)	1,087.8	(35.8)	1,439.0	(37.3)
Nuclear	26.8	(1.3)	36.6	(1.2)	58.5	(1.5)
Renewables	182.9	(8.6)	239.0	(7.9)	324.4	(8.4)
Total	2,134.1	(100.0)	3,036.6	(100.0)	3,853.7	(100.0)

Note: Calculated from data in USEIA (1995).

[a] Forecasts are for the "reference case" scenario, which is between the high- and low-growth scenarios.

composition is not expected to change significantly. Although oil and natural gas use will almost double during this period, their share in total fuel used is expected to increase only slightly. The use of oil and natural gas by developing countries is forecast to grow by 2 and 1.8%, respectively, while that of renewables is projected to increase by 2.5% on an annual average basis. Natural gas and coal use is expected to increase by 1.6% per year. All growth rates are higher than those of developed countries, except for natural gas use (which in developed countries is expected to increase by 2.2% per year during the 20-year period, reflecting, in part, concerns over the environment). The increase in renewable fuel use worldwide and its share of total fuel use reflects an expected increase in hydroelectric power over the next two decades. Although not reflected in these figures due to a lack of data, non-commercial use of renewable energy from plant and animal sources in developing countries is expected to decline as economic growth occurs and lifestyles change.

While energy use will increase considerably in both developed and developing areas, Table 12.3 shows that the bulk of the increase is expected to originate in developing countries. It is important to note, however, that energy use in developing countries will still be not even half that used by developed countries by 2010.

The increased use of energy, together with the noted relatively unchanged composition in fuel type, has negative implications for sustainable development. The dominance of fossil fuels in overall energy use is expected to continue, although a slightly greater reliance on natural gas by developed countries is noted. The dependence on coal, which has the most harmful effect on the environment, will decline in both developed and developing countries. Although this is a positive trend in terms of environmental sustainability, coal use will still increase in absolute terms during the 20-year period and most of this increase will be due to greater use in developing countries.

This and other similar forecasts, if correct, bode ill for the state of the environment in the coming century. Total carbon emissions would increase from 6 billion metric tons in 1990 to over 8 billion in 2010 (USEIA, 1995). Developing countries, which accounts for 29% of total carbon emissions in 1990, are expected to contribute 38% by the year 2010. Developing countries are also expected to increase their carbon emissions by almost 4% per year by 2010, while developed countries will experience an increase of less than 1%. Other forecasts are similar, with predictions of developing countries' contribution to world carbon emissions at 36% in 2005 and 50% by 2020 (USEIA, 1995; United Nations, 1992). It is also noteworthy that while carbon emissions are growing at a rate lower than energy consumption in developed countries, the opposite is true in developing countries.

Changing the fuel mix has been a successful means of reducing local pollution in developed countries. Natural gas has been substituted for coal and oil to a significant extent, with positive environmental effects. Table 12.3 shows that the coal supplied in developing countries is expected to remain about the same in absolute terms and decline in importance as a fuel by 2010. In developing countries, however, the amount of coal supplied will increase substantially in absolute terms and decline only slightly in overall importance. Given the importance of coal as an energy source, it is important for clean coal technologies to be adopted, especially in developing countries. Reliance on natural gas, on the other hand, is expected to increase significantly in developed countries (continuing a trend witnessed since the 1970s) but only moderately in developing countries. The success of developed countries in switching to a more environmentally friendly fuel has been accomplished by tax incentives combined with private financing for sophisticated transportation systems. The success of such measures requires political will and the ability to implement selective tax schemes, as well as access to large sums of capital.

Recently, many developing countries have adopted policies of gradually lifting direct and hidden subsidies on energy products so that prices are more in line with opportunity costs. This seems to become more politically feasible as market-oriented policies are adopted in the wider economy. A further step in promoting sustainable development is to encourage a pricing system that accurately reflects the cost of pollutants of each energy source, although currently this may not be politically feasible in developing countries.

It should be noted that a significant portion of the energy supply in developing countries is non-commercial and often renewable. Data on this type of energy are difficult to gather on a systematic basis and are often not included in energy analyses. Indeed, Table 12.3 does not include non-commercial energy sources, and its data on renewable sources are underestimates. However, renewable sources are used in developing countries, particularly in rural areas where people rely on solar energy, biomass, biogas, and wind energy. A significant portion of the population in developing countries is not connected to an electricity grid, relying instead on a combination of renewable sources and inefficient generators which require expensive batteries or fossil fuel. Although often dismissed as far less efficient than fossil fuels and other conventional energy sources, renewable energy sources are being reconsidered, especially for rural areas. One recent study concludes that if the environment- and health-related costs of burning fossil fuels are included in their price, renewable sources are actually less expensive than fossil fuels per unit of energy delivered (Hohmeyer, 1992). Another study notes that the widespread adoption of renewable energy sources is hampered by a lack of institutional frameworks; the infrastructure available in most developing countries favors a centralized, capital-intensive energy supply (Hulscher and Hommes, 1992). In the long term, renewable energy may be a solution to present-day environmental problems caused by the use of fossil fuels.

Renewable energy also includes hydroelectric power, which, although desirable when considering global environmental effects, may produce environmental degradation at the local level. Table 12.3 shows a significant increase in the global use of renewable energy, which is mainly a reflection of hydroelectricity in both developed and developing countries. The production of electricity in developing countries has to a large extent been the responsibility of governments, although with economic restructuring and privatization programs there is a move toward private sector involvement, coupled with a deregulation of prices. This is an important institutional change, because the private sector has a powerful incentive to conserve energy and reduce wasteful production practices, while higher prices encourage a reduction of waste by the final consumers. However, sustainable development requires that the private sector include environmental degradation from production in its costs. This will be difficult to accomplish within the current political and institutional frameworks in developing countries.

Full cost pricing of energy, or any other basic necessity, is not without adverse social implications. Developing countries in particular are faced with the dilemma of how to provide essential goods at prices affordable to relatively poor populations. While pricing policies may be useful in allocating energy resources, including the share of each fuel in total energy supply, they may impose hardships on those least able to afford them. However, to achieve environmental sustainability, welfare goals, including those of income growth and distribution, need to be addressed with non-energy policies. Actually, this issue is an important political consideration for developing countries that implement policies aimed at promoting energy efficiency and changing the supply mix of energy products. While governments may recognize the importance of pricing policies, they often opt for a gradual approach to lifting subsidies in an effort to ease the hardships of higher prices.

Egypt

Egypt is faced with myriad problems that impede sustainable development. Its population has been growing by approximately 2% per year since 1985, while per-capita income (about $670 per year) has grown by less. Arable land constitutes only 5.5% of the country's approximately 1 million km^2, and the population density is over 700 people per square kilometer of arable land. Egypt is faced with a number of environmental problems, the most notorious being the air pollution in Cairo. Cairo's environmental problems stem from overcrowded conditions, traffic congestion, industrial emissions, incineration of garbage, and dust from natural sources. In 1990, smoke emissions from vehicles in Cairo were estimated at 1200 tons or seven times that recorded in 1970 (United Nations Environment Program and World Health Organization, 1994). Nearby industrial complexes produce ceramics, metals, glass, bricks, textiles, and plastic. They are powered by electricity generation plants fueled by heavy fuel oil and natural gas. Current levels of particulate matter are far in excess of World Health Organization guidelines. Nitrogen oxide levels were measured in 1991 at 0.2 ppm in downtown Cairo (Neamatalla, 1991). Other problems faced by Egypt include air pollution in Alexandria and other large cities and towns and water pollution from wastage of many industries, including oil refining.

Energy consumption has been growing by 3.4% per year since the late 1970s, reflecting economic growth, growth in urban areas, growth in the number of vehicles, and a significant expansion of the road system. Egypt's energy needs have been satisfied with hydrocarbon fuels, mainly oil and natural gas, and hydroelectric power. Table 12.4 shows Egypt's energy consumption by type of fuel. Oil use accounts for over 65% of total energy used, while gas is second in importance, accounting for 29%. Although Egypt's Aswan Dam is well known for its regulation of flood waters, hydroelectricity accounts for only 3% of Egypt's total energy consumption.

Egypt's energy consumption pattern is influenced by resource endowment, which is mainly in the form of oil and gas. Egypt has encouraged exploration activities during the past 20 years and has successfully exploited newly discovered oil and gas fields. Oil reserves increased from less than 3 billion barrels in

Table 12.4   Energy Consumption by Fuel Type in Egypt, 1980–92 (in Thousand Metric Tons of Oil Equivalent, Except Per Capita)

	1970	1980	1985	1990	1992
Per capita[a]	181	358	481	488	493
Coal	318	471	729	853	659
Oil	5,171	12,540	17,583	17,562	17,711
Gas	78	951	3,340	6,320	7 825
Hydroelectricity	398	812	913	858	834
Total	5,965	14,774	22,565	25,593	27,029

[a] Kilograms of oil equivalent.

Source: United Nations (1970, 1980, 1985, 1990, 1992).

1980 to more than 6 billion by 1994, and oil production increased from 550,000 barrels per day to almost 900,000 during the same period. The incentive for exploration and development activities has come not only from a rapid increase in domestic production but also from the prospect of international oil sales and the associated hard-currency earnings. Although not a major player in terms of the international oil market in general, Egypt's oil revenues have had a positive impact on its development potential. The country's oil revenues increased from US$164 million in 1975 to a peak of US$2.6 billion in 1983 and, although fluctuating annually due to changes in oil prices, amounted to US$1.2 billion in 1994.

In fact, efforts to maintain significant hard-currency earnings from oil sales have been a major focus of Egypt's energy policy. In the face of rising domestic demand and rather limited reserves, Egypt found that domestic energy consumption growth was lowering the amount of oil for sale on the international markets. In the early 1990s, the country adopted a policy of substituting natural gas for petroleum products wherever possible and has been fairly successful in this endeavor. The exploitation of natural gas had been lagging due to unfavorable terms for oil and gas companies, which often ignored gas discoveries. A revised natural gas clause was devised, with economic incentives to companies to search for and exploit natural gas reserves. As a result, natural gas reserves have increased considerably, from 351 billion cubic meters (bcm) in 1990 to 436 bcm in 1993 to 546 bcm in 1995. A recent estimate indicates that undiscovered reserves could amount to one trillion cubic meters. In addition, infrastructure development, most notably of the gas grid in Cairo, which pipes gas directly to households, was undertaken. As a result, natural gas use increased considerably in Egypt and continues to grow. Although it is not reflected in Table 12.4, gas use in Egypt increased by more than 14% in 1993 and by almost 9% in 1994. Sixty-four percent of natural gas is used for electricity generation and 14% for the production of fertilizer, with the remainder going to domestic low-propane gas and other industrial uses.

Along with efforts to increase the supply of natural gas, Egypt has undertaken demand-oriented policies aimed at substituting gas for oil and oil products. Some electricity-generating plants have switched to gas as part of this effort, and Egypt has been gradually increasing the domestic prices of petroleum products and electricity. Domestic prices of petroleum products reached international levels in 1994, while electricity prices were 80% of world prices in 1994 and reached international levels in 1995. Although faced with initial political opposition, this increase in prices is an important part of Egypt's overall energy policy and plays an important role in the sustainable development of the country.

While the emphasis on gas as a substitute for oil and oil products has been made with hard-currency earnings as a major goal, the switch in consumption has been favorable to the local environment. Natural gas, although a contributor to the global environmental degradation, is a much more desirable fuel than oil and coal in terms of localized environmental degradation. It is difficult to measure the benefits of natural gas in this particular case, since many other factors have contributed to Egypt's environment problem. Generally, however, natural gas use is associated with almost no sulfur dioxide emissions in all consuming sectors, about half of the nitrogen oxide emissions from oil in the domestic and industrial sectors, and about two-thirds of nitrogen oxide emissions in the electricity-generating sector (International Energy Agency, 1988).

To achieve the goal of sustainable development, a complex mix of policies must be undertaken, ranging from direct regulation of the transport and industry sectors to a long-term process of educating the public on conservative energy use. Along with its forward-looking energy policies, Egypt has undertaken steps toward privatization as part of an overall macroeconomic restructuring program. Encouraging a newly emerging private sector to adopt the most environment-friendly production process in terms of energy use is one of the biggest challenges facing Egypt in terms of sustainable development. Egypt is typical of developing countries in that the institutional infrastructure may not be able to enforce regulations or ensure the success of a complicated tradeable permit scheme. Positive incentives, such as tax breaks for environment-friendly actions on the part of industry, also require a well-developed public finance system. The regulation of the transportation sector in terms of emissions standards may be easier to enforce, at least in terms of new vehicles sold. A review of the vehicles in the public transportation fleet with a view toward establishing a self-imposed emission standard might be a good place to start.

Policies aimed at rural development offer another solution to the growing energy consumption in urban populations (Jones, 1991). Egypt's environmental problems are particularly serious in its urban areas, especially Cairo and Alexandria. Attempts to relocate the population from Cairo have not been overwhelmingly successful, as people tend to commute longer distances rather than relocate. Actually, Cairo's traffic problem is compounded by a daily influx and exit of workers amounting to approximately two million people per day. Rural development programs, which would improve the economic prospects of village and small-town dwellers, would reduce the temporary and permanent movement of Egypt's population to already congested urban areas.

Long-term sustainable development also requires population policies as well as efforts aimed at raising the awareness of the population about environment-related issues, including energy use. Although population growth has been declining in developing countries including Egypt, its large population poses an obstacle to achieving sustainable development goals, especially with regard to resource use. The problem is widely recognized within the country and, although a culturally sensitive issue, has been addressed with some success by policymakers. Population growth has declined to around 2% in the 1990s from 2.5% a decade earlier. Education through the university level is free in Egypt, and the education of the general public has been a significant achievement of the government since the 1952 revolution. However, many children from impoverished families do not attend school or leave school early due to economic pressures. The education of girls is given less priority than education for boys within the family in many cases. The education of girls is considered particularly important for sustainable development since they will, when they become adults, make many of the household decisions concerning the use of resources, including energy. Egypt, with its extensive education system in place, could introduce an energy/environment awareness program into the curriculum rather easily. Other education programs aimed at adults may have a more immediate impact.

South Africa

South Africa's energy profile has been affected not only by its resource endowment but by its political and institutional structure, which (until recently) rigidly divided society by race. Today, although South Africa is now a democracy with all racial segregation laws abolished, the population's lifestyles, income levels, employment patterns, and energy use are still influenced by past practices. Although considerable change has occurred socially and economically as well as politically, it is a long-term process, and current and future resource use reflects living patterns that are the results of apartheid. On the supply side, coal is the major energy source, and its use was encouraged by high oil prices in the 1970s and early 1980s and by the reaction to increasing international isolation and the threat of economic sanctions later in the 1980s, which led the government to consider coal as vital to the security of the nation.

From an environmental perspective, South Africa's reliance on coal is the least desirable choice, although the country's coal has a relatively low sulfur content. The country's coal reserves amount to over 55 billion tons, which account for 91% of all known reserves on the African continent (British Petroleum, 1989–95). South Africa produced 195 million tons of coal in 1994, which was mainly used for the production of electricity and synthetic petroleum products. However, the country's oil and gas reserves are very low, 40 million barrels and 27 bcm, respectively, a fact reflected in the consumption patterns shown in Table 12.5. Table 12.5 does not show that coal contributes almost 80% of the primary fuel used in South Africa, since a large proportion of electricity-generating plants utilize coal and about one third of South Africa's petroleum products are synthetic fuels produced from coal (van Horen et al., 1993). South

Table 12.5 South African Energy Consumption by Fuel Type and Consuming Sector, 1994 (in Thousand Metric Tons of Oil Equivalent)

	Coal	Petroleum products[a]	Gas[b]	Electricity	Total by sector	Percentage by sector
Industry	12,438	2,271	462	7,735	22,906	53.9
Transport	11	10,706	—	307	11,024	25.9
Residential	989	482	8	2,107	3,586	8.4
Public/commercial	884	199	15	1,602	2,700	6.4
Agriculture	104	1,159	—	347	1,610	3.8
Other/non-energy	—	693	—	—	693	1.6
Total by source	14,426	15,510	485	12,098	42,519	100.0
Percentage by source	33.9	36.5	1.1	28.5	100.0	

[a] Including coal-based synthetic fuels.
[b] Coal-based gas.

Source: International Energy Agency (1988, 1992).

Africa has the only nuclear power plant in Africa, which provided 2.6 million TOE in 1994. South Africa's rural population consumes non-commercial energy (not included in Table 12.5), mostly biomass such as firewood, which accounts for approximately 6% of total energy used.

Current energy policy is directed toward restructuring a partially state-owned energy industry supported by subsidies, as well as providing for the energy needs of the poor. High oil prices in the 1970s and early 1980s and a concern over growing international isolation under which sanctions on oil imports could have been enacted prompted investment in an extensive synthetic fuel industry with a production capacity of almost 200,000 barrels per day. As oil prices fell and remained low in the mid 1980s and early 1990s, South Africa regulated oil product prices and directly subsidized the high-cost synthetic fuel industry. Today the subsidies to one producer, Sasol, amount to one billion rand per year (equivalent to US$273 million). This subsidy and the deregulation of product prices are currently topics of discussion between government and industry officials.

The electricity industry is also under scrutiny in the current consideration of ways and means of providing electricity to the two-thirds of the population currently without access in both rural and urban areas, despite an excess generating capacity. Although South Africa produces more than half the electricity sold on the continent of Africa, its distribution within the country has been heavily influenced by institutional factors as well as the historically high degree of income inequality. Electricity is provided by local authorities which catered to localities divided according to race. Those serving areas restricted to whites under apartheid were well staffed with qualified personnel and established electricity networks, well-funded capital development funds, and a diversified customer base with few demands for new connections, while those serving areas restricted to blacks under apartheid were generally understaffed, with underdeveloped administrative capacity, limited access to capital, high demand for new connections, and a customer base of mainly low-income residential users (van Horen et al., 1993). Local elections held in 1995, aimed at integrating the local governments of formerly white and black areas, will help to solve some of these institutional problems. An important part of the political resistance during apartheid took the form of rent and service boycotts, including nonpayment of electricity bills. Power cutoffs and high reconnection costs became a political issue, and many residents became used to not paying for services, which poses an additional institutional problem when planning for the financial aspects of the considerable investment in electricity infrastructure required for new connections.

The inadequacy of the system in providing electricity has had adverse effects on the environment, as urban dwellers use coal, kerosene, candles, and liquefied petroleum gas for lighting, cooking, and heating. Coal use varies according to income and proximity to coal fields, with low-income families and those living closer to coal fields using greater amounts than otherwise. The price of coal includes transportation costs which factor into the final price, and low-income households are very responsive to variations in price. The coal used by these households is generally of the lowest quality and generates high emissions when burned. Although it accounts for only 6.8% of total coal use, the residential use of coal is a major contributor to localized pollution. The high ash content of South Africa's coal produces significant concentrations of suspended particulates in formerly black areas, especially during the winter as the demand for heating fuel rises. The pollution from coal is immediately apparent to anyone in these areas, particularly during the early evening when dinner is being prepared. Studies and surveys of the formerly black areas have shown that there is considerable substitution of fuels in areas where partial electrification is present, and the totals of suspended particulates remain high unless complete electrification of an urban area is accomplished. Low-income urban dwellers, especially children, have had adverse health problems as a result of suspended particulates and other pollutants from coal, such as sulfur dioxide, carbon monoxide, nitrogen dioxide, polycyclic aromatic hydrocarbons, and benzopyrene (Terblanche et al., 1992, 1995).

More generally, coal use by all consuming sectors causes pollution in South Africa. Total greenhouse gases, for instance, account for 2% of global greenhouse gases, although South Africa's population is only 0.7% of the world population (van Horen et al., 1993). Its carbon dioxide emissions amount to 279 million tons, or over 7.1 tons per person compared to 1 ton per person in Africa as a whole and 4.2 tons per person worldwide (World Resources Institute, 1994). Eighty-seven percent of the country's carbon dioxide emissions result from burning coal. Electricity-generating plants have not invested in environmentally friendly technologies such as desulfurization technologies, and as the electrification of the formerly black areas progresses and new capacity is added, increased emissions of sulfur dioxide, nitrogen oxides, and carbon dioxide will occur. Satisfaction of additional demand should be accompanied by appropriate environmental planning. Officials are also in a position to encourage the use of energy-efficient appliances in households; since a large number of households currently have no electrical appliances, incentives for their production and purchase can be very effective. Unlike most other developing countries, a well-developed institutional infrastructure exists in South Africa which makes regulation more likely to be effective. An established public finance system along with a developed capital market make a system of tradeable pollution permits more feasible than in most developing countries.

Current trends toward urbanization are expected to continue, and by 2020 South Africa is expected to be characterized by an urban population comprising 66% of its total, compared with 51% today (United Nations, 1995). This trend implies a greater potential demand for commercial energy and highlights the need to satisfy urban energy demand without degrading the environment. Rural energy needs must also be met and, like Egypt, South Africa's environmental problems will be helped by a rural development program, especially since the mostly rural former black homelands were systematically ignored under apartheid. South Africa's recent political accomplishments are laudable and recognized worldwide, but the new government faces economic challenges in its efforts to improve the living standards of the majority of the population. It is important for environmental considerations to be included in any assessment of overall living conditions.

Although South Africa currently relies mostly on its coal reserves, potential exists to utilize natural gas and hydroelectric power provided by nearby countries. As South Africa develops, its economic and trade relations with neighboring countries are expected to grow, with economic benefits spilling over to the entire region, including the possibility of developing gas and electricity networks. There are significant natural gas fields in Namibia and Mozambique which could provide a cleaner burning fuel for South Africa. Currently, natural gas consumption is very limited, as its supplies are small and almost exhausted. Developing the well-known hydroelectric potential of sub-Saharan Africa could also eventually be used to satisfy South Africa's growing energy needs. South Africa is also encouraging exploration for oil and gas in areas considered to have potential fields, mostly in offshore areas. The long-term potential of South Africa diversifying its energy sources away from a dependence on coal is favorable from an environmental perspective.

China

China consumes over 9% of the total energy consumed worldwide and accounts for 11% of the total carbon dioxide emissions from industrial processes. China's significance at the global level from an environmental perspective also lies in its potential contribution to carbon dioxide and other harmful atmospheric pollutants. Although per-capita income is low ($360 in 1993), China's population is the largest in the world at 1.2 billion. Economic growth has been high, with the increase in real per-capita income measured at over 6.6% per year from 1974 to 1990 and over 11% since then (World Bank, 1995a). China's growth has been stimulated by a liberalized economic policy introduced gradually during the 1980s and 1990s and an economic boom, especially in the southeast region of the country. Current forecasts indicate that economic expansion will continue well into the 21st century. This, coupled with the increasing population which is expected to reach 1.4 billion by 2010, foretells higher energy consumption in absolute and per-capita terms and higher emission levels as a result.

China's energy consumption increased significantly from 1980 to 1994, by about an average of 3.2% per year, mainly due to large increases in the use of coal and oil. China also began using nuclear power in 1991. As with other developing countries, China's energy consumption pattern has been influenced by resource endowments as well as institutional, political, and economic factors. The heavy reliance on coal shown in Table 12.6 reflects China's significant coal reserves, totaling 115 billion tons or 11% of the world's total. China's consumption pattern also reflects its past policy of encouraging oil exports during the 1970s and early 1980s, when prices were high, and a centrally planned investment decision that neglected the development of natural gas reserves. Until recently, energy prices were controlled at levels below production costs. Table 12.6 covers commercially traded fuels only, but the use of traditional fuels such as biomass and biogas is widespread in rural areas, accounting for 7% of total consumption if included in totals (World Resources Institute, 1994).

Undertaking sustainable development policies in China requires a serious reconsideration of its heavy reliance on coal. China has taken steps to diversify its energy sources and, as can be seen in Table 12.6, its use of oil has increased considerably. Although China's use of oil has almost doubled in over 14 years, the relative importance of oil in overall energy use has increased only moder-

Table 12.6 Commercial Energy Consumption by Fuel Type in China, 1980–94 (in Millions of Metric Tons of Oil Equivalent)

	1980	*1985*	*1990*	*1994*
Coal	403.3	509.4	518.9	572.0
Oil	88.0	88.9	113.4	144.1
Nuclear	0	0	0	3.1
Natural gas	11.7	11.5	13.2	14.9
Hydroelectricity	15.0	25.8	31.5	14.5
Total	518.0	635.6	677.0	748.6

Source: British Petroleum, 1989–95.

ately, from 17 to 19%. Coal is an important fuel for the production of electricity in China, and in 1993, 65% of the total installed capacity was coal fired (Li and Dorien, 1995). Coal demand is expected to increase by 76% by 2010, but the increase in the use of coal for generating electricity is projected to increase by 138% (Wu and Li, 1995). As demand for electricity grows and as it is vital for economic growth, China is unlikely to curb its use. However, there is still considerable room for improved efficiency and adoption of clean coal technologies in China, and this can have considerable impact on future emission levels. Like South Africa, China utilizes coal for residential heating and cooking. Although a small portion of total coal consumption, residential use of coal totaled 167 million tons in 1993. While this is expected to decrease as urban areas become electrified, it is a significant source of pollution in urban areas, as well as an inefficient use of energy. One recent study notes that coal stoves with thermal efficiencies of only 15 to 30% were used in more than 60% of the urban households. The switch by these households to electricity produced by clean coal technologies may be expected to have a considerable positive effect on the environment.

Hydroelectricity has unrivaled potential in China, and the government is investing in this sector to satisfy increasing energy demand and as an environment-friendly source of energy. Installed capacity is expected to increase from 49 gigawatts in 1994 to 125 gigawatts in 2010, when it will account for 24% of total electricity produced. The Three Gorges Project is a good example of the unsteady equilibrium governments try to achieve between satisfying energy demand to promote economic growth and safeguarding the environment. This project has come under considerable scrutiny and criticism worldwide due to the havoc it will cause to the local environment. Under construction since 1993, the Three Gorges Dam will be the world's largest hydroelectric facility, with a capacity of 18.2 gigawatts, and will cost about $11 billion. Its yearly production will be equivalent to that produced with 40 million tonnes of coal. Criticism of the project centers around its social costs—1.1 million people are being resettled—as well as the environmental harm to local ecosystems, including endangered species, and the adverse effects of possible silt buildup.

Policies regarding oil have changed from export for hard currency to increasing use at the domestic level. The policy of maximizing exports was halted in 1986 when international prices fell. China became a net oil importer in 1992, and imports amount to about 25% of its total oil consumption; expectations are that China will need to import 2 million barrels per day by 2005 (Ryan and McKenzie, 1995). Current demand is about 3 million barrels per day. Growth in domestic production is currently limited by depleting reserves, and China is inviting foreign oil companies to explore for oil in both onshore and offshore areas. China also faces a refinery capacity shortage which is especially acute in the southeast region of the country, where demand has been growing rapidly as the regional economy has expanded.

Natural gas potential exists in China, and since its use has the least adverse environmental impact, its development would be a positive contribution to China's efforts at promoting sustainable development. Natural gas currently accounts for only 2% of total energy consumption, and the industrial sector is the major consumer. The chemical and oil and gas industries consumed most of the gas, with residential use only 11% of the total (Wu and Li, 1995). At present, China's reserves of natural gas amount to 1.7 trillion cubic meters or about 1% of the world's total reserves (*Oil and Gas Journal,* December 26, 1994, p. 44). However, geologic formations lead some to believe that China's natural gas reserves may be significantly higher. The gas industry has stagnated for a number of reasons, including official neglect, market fragmentation, lack of transportation facilities, lack of exploration and production, low investment, and prices controlled at very low levels (Wu and Li, 1995). The government is planning to increase exploration and production activities as well as to utilize price reforms to stimulate gas production. Forecasts of higher natural gas demand indicate that China may become a gas importer, potentially importing 43 million cubic meters per day by 2010; despite recent price increases to residential users, it is still less than imported liquefied natural gas (Ryan and McKenzie, 1995).

Price reforms in the energy industry have the potential to influence the energy consumption mix as well as to encourage conservation. China is undergoing significant change in its economic structure, including price reform in general (Wu and Li, 1995). Revising energy prices to reflect production costs is also a feature in its efforts to supply enough energy to fuel its expanding economy. Major reforms in the pricing of oil were initiated in 1994, eliminating a two-tiered pricing formula under which most oil was sold at very low prices. Prices of oil and oil products are now higher but still below market prices. Coal price controls were eliminated in 1994, although prices had varied by location and time according to transportation and production costs. Subsidies on transportation, however, still enable coal to be sold at lower prices than otherwise. As mentioned above, natural gas prices are under reform as part of China's efforts to stimulate supply.

Electricity prices have been increasing since 1985 but are still the lowest in Asia at US2.5 cents United States per kilowatt-hour. (As a comparison, prices in the United States average 6.9 cents per kilowatt-hour.) Price reform of energy, although a start, remains an area in which sustainable development can be encouraged. China faces a dilemma here as well, since low energy prices can be a form of welfare, and per-capita income is very low. However, as noted above, pricing energy according to its production costs, including costs to the environment, remains one of the most effective tools available for promoting sustainable development in developing countries.

Efforts at improving energy efficiency are also under way, yet considerable room for improvement also exists in this area. The residential use of coal as a fuel for heating has been mentioned, but this consists mostly of raw coal. The introduction of molded coal would improve its thermal efficiency, and the addition of a vulcanized agent could decrease sulfur dioxide emissions by half (Geping, 1992). Aside from the use of inefficient coal-burning stoves, China also has a large number of small and medium-sized coal-fired electricity-generation facilities, over 60% of which had installed capacities of less than 200 megawatts, which are generally considered inefficient. Larger facilities are now given priority, and their increased use will ensure that energy is used efficiently and have a lower negative impact on the environment, but it will take some time before the smaller inefficient facilities can be phased out. Investments have been made in energy conservation programs in other industries as well, with apparent positive effects. China has made progress in delinking economic growth and energy use, although there is some debate as to how this has been achieved (Sinton and Levine, 1994; Zang, 1995).

The institutional structure related to the energy industry has also undergone change which should lead to a more efficient allocation of resources. Petroleum exchanges have been established on the country's nascent stock exchanges, private oil companies have emerged as a result of encouragement to the private sector, tax reform has been applied to state-owned refineries and foreign investment, and participation in the industry is being encouraged. The management of coal mines is being decentralized, with incentives introduced to improve efficiency. In addition, other economic liberalization measures, such as issuing of trade licenses and easing import quotas in some cases, have affected the energy industry.

Comparison of Countries and Conclusion

Given that environmental degradation resulting from energy use is minimal in most developing countries, notwithstanding some notorious urban areas, and that contributions to global emissions are low, especially on a per-capita basis, as developing countries embark on economic development programs they are in a good position to enact measures to safeguard the environment. On the other hand, there is a need for institutional reform and strengthening, as well as for a political will to regulate and control activities that cause environmental degradation. Often measures successful in developed countries are not appropriate for developing countries, and solutions to environmental problems unique to the developing world have not been found or tested. Long-term solutions, however, may coincide with economic and social development measures planned to accomplish other goals, such as education.

Egypt, South Africa, and China have all embarked on policies aimed at raising the standard of living, and they are all expected to increase energy consumption. Recent income levels, energy consumption levels, and the fuel mix of each country are shown in Table 12.7. Undertaking policies aimed at sustainable development requires that countries raise income levels without

Table 12.7 Comparison of GDP, Energy Consumption, and Fuel Mix in Egypt, South Africa, and China

	Egypt	South Africa	China
Real GDP ($bn)[a]	40	85	430
Per-capita GDP ($)[a]	670	2150	360
Energy use (million TOE)[b]	27	43	749
Per-capita energy use (kg OE)[c]	493	1000	600
Fuel mix (%):			
Oil	65.5	19.8	19.2
Gas	29.0	0	2.0
Coal	2.4	77.3	76.0
Hydro	3.1	2.7	1.9
Nuclear	0	0.2	0.4

[a] 1993 data based on 1987 prices.
[b] Million metric tons of oil equivalent.
[c] Kilograms of oil equivalent.

Source: World Bank, 1995b; United Nations, 1995; British Petroleum, 1995; and Tables 12.1–6.

adverse effects on the environment. With regard to energy, a combination of policies is needed to control demand and to encourage the supply of fuels that are associated with the least harmful environmental effect. Fuel use is often influenced by supplies available within the borders of a country, as has been the case for the three countries considered.

Egypt has been fairly successful in changing its fuel mix from oil to gas through pricing policies aimed at discouraging oil use, infrastructure development favoring gas use, and pricing incentives to suppliers. The policy undertaken to reduce subsidies on energy products is expected to result in greater energy efficiency as well. Egypt's move toward a greater reliance on the private sector in the overall production of goods and services should also result in greater energy efficiency. Additional incentives, however, are needed in the industrial and transportation sectors, since they are major sources of urban pollution. Specific measures targeted at the transportation industry may be recommended to encourage rational energy use. Egypt might want to consider emission control standards and improvement of the mass transportation system in terms of energy efficiency as well as the scope of its provision.

To achieve sustainability, efforts must be made to internalize the costs of using the environment as a free good. This is perhaps the most difficult task for all countries, but it is essential for sustainable development. As standards of living improve and as institutional development occurs, Egypt will be in a better position to undertake measures to force users of the environment to pay the full cost of degradation. Also, longer term social policies with energy implications, such as population policies, should be undertaken with sustainable development in mind.

Although South Africa's income levels and energy consumption levels are much higher than those in Egypt, as shown in Table 12.7, South Africa is also concerned with economic development to eliminate the wide disparity in income among social groups and the widespread poverty in the country. To an extent, South Africa's pollution problems are those typical of a developed country, and although its commitment to raising living standards will mean energy consumption in the future, it has the advantage of a well-developed institutional infrastructure with which to implement sustainable development measures. The major problem of relying heavily on coal is under reconsideration, as is the state support of the energy industry. Diversification of energy sources, though only begun, is important, especially in light of the expected increases in demand. Sustainable development policies should include special attention to providing electricity, using clean coal technology and hydroelectricity facilities, to those areas that currently burn raw coal for cooking and heating. Attention to rural development is needed to counteract increasing urbanization. South Africa is in the best position of the three countries, and among developing countries in general, to implement measures aimed at internalizing the costs of the environment, and such measures should be seriously considered.

China's per-capita income is the lowest of the three countries considered in this chapter and is low by global standards as well. On the other hand, China consumes the most energy, as shown in Table 12.7, and as its economy grows, this will have a significant impact on world energy demand and emission levels. China's recent policies reflect a concern about environmental degradation resulting from energy use, but the country is hesitant to accept lower growth levels, given its very low per-capita income level. Initial efforts to deregulate energy prices are a step toward full cost pricing of energy, but such a policy must be consistent for all fuels and remain unchanged. China has recognized that it must adopt cleaner coal-burning technologies and has made efforts toward encouraging more efficient coal-burning electricity-generating facilities. China's fuel mix is heavily weighted toward coal, and sustainable development policies should include a diversification of fuel types. Its current efforts at increasing hydroelectric power will help to prevent additional emissions from carbon fuels. China's current early stage of development actually allows it a degree of flexibility in choosing a sustainable development course. For instance, rural development plans can rely in part on renewable energy technologies already utilized in these areas. Measures aimed at the transportation sector are not as hindered by existing energy-inefficient systems already in place, as in the case in Egypt for example. Promoting energy efficiency with broad-based pricing policies, as well as specific policies targeted toward specific problem users, should be an important part of China's strategy. China's ability to internalize the costs of degrading the environment is currently minimal, but as its economy grows, its financial and institutional infrastructures will develop and such measures may be feasible in the future.

Conclusions

The environment tends to be a normal good (i.e., it is valued more as the standard of living rises). Yet, each country considered here has taken the environmental aspects of energy-related decisions into consideration to some degree, especially those related to degradation at the local level. Each country faces difficult choices between promoting economic growth and safeguarding the environment. Changes in the production process or the adoption of environment-friendly technologies, however, are often very expensive for developing countries, and require capital investments beyond the means of most of them. China's gradual approach to the adoption of clean coal technologies is largely a response to these constraints. At any rate, the differences among the three countries highlight the importance of diversified strategies tailored to suit the specific circumstances in different developing regions.

As developing countries turn more to a market system for the allocation of resources, it is an appropriate time for the adoption of market-oriented measures to allocate the once overlooked resource of a clean environment. As an economy develops and the institutional features of a market economy are strengthened, market-oriented solutions such as tradeable permits become more feasible. However, most developing countries will not be in a position to utilize such measures effectively for some time.

Global environment problems are less of a priority for developing countries more concerned with raising standards of living above a subsistence level and local environmental problems. Developing countries point to the developed countries as the major source of global environmental problems and as better able financially to address these problems. As emissions from developing countries which affect the global environment are growing, developed countries may find that aid and technical assistance programs to reduce these emissions would be beneficial not only to the receiving countries but also to the rest of the world, including themselves.

References

Abdalla, K.L. (1994) Energy policies for sustainable development in developing countries. *Energy Policy,* 22(1): 29–36.

British Petroleum (1989–95) *BP Statistical Review of World Energy.* Ashdown Press, London, U.K.

Geping, Q. (1992) China's dual-thrust energy strategy. *Natural Resources Forum,* 16(1): 27–31.

Goldemberg, J., Johansson, T., Reddy, A., and Williams, R. (1986) *Energy for Development.* World Resources Institute, Washington, D.C.

Hohmeyer, O. (1992) Renewables and the full costs of energy. *Energy Policy,* 20(4): 365–375.

Hulscher, W.S. and Hommes, E.W. (1992) Energy for sustainable rural development. *Energy Policy,* 20(6): 527–532.

International Energy Agency (1988) *Emission Controls in Electricity Generation and Industry.* Organization for Economic Cooperation and Development, Paris, France.

International Energy Agency (1992) *Global Energy: The Changing Outlook.* Organization for Economic Cooperation and Development, Paris, France.

Jones, D.W. (1991) How urbanization affects energy use in developing countries. *Energy Policy,* 19(7): 621–630.

Levine, M.D. and Meyers, S. (1992) The contribution of energy efficiency to sustainable development. *Natural Resources Forum,* 16(1): 19–26.

Li, B. and Dorien, J.P. (1995) Change in China's power sector. *Energy Policy,* 23(7): 619–626.

Neamatalla, M.S. (1991) Urbanization and the Environment, Case Study of Egypt. United Nations Report E/ESCWA/ENVHS/1991/CRP/3.

Ryan, D. and McKenzie, W. (1995) More help needed to boost output. *Petroleum Economist,* 62(6): 14–16.

Sinton, J.E. and Levine, M.D. (1994) Changing energy intensity in Chinese industry. *Energy Policy,* 22(3): 239–255.

Terblanche, A.P.S. et al. (1992) Preliminary results of exposure measurements and health effects of the Vaal Triangle air pollution study. *South African Medical Journal,* 81: 550–556.

Terblanche, A.P.S., Nel, C.M.E., and Tosen, G.R. (1995) Respiratory health impacts of three electrification scenarios in South Africa. *Journal of Energy in Southern Africa,* 6(2): 93–96.

United Nations (1970, 1980, 1985, 1990, 1992) *Energy Statistics Yearbook.* United Nations, New York, NY.

United Nations (1995) World Urbanization Prospects, The 1994 Revision. Report ST/ESA/SER.A/150. United Nations, New York, NY.

United Nations Environment Program and World Health Organization (1994) Cairo: unbridled dust. *Environment,* 36(2): 30.

USEIA (1995) International Energy Outlook (1995). United States Energy Information Agency, Washington, D.C.

van Horen, C., Eberhard, A., Trollip, H., and Thorne, S. (1993) Energy, environment and urban poverty in South Africa. *Energy Policy,* May: 623–639.

World Bank (1995a) *Global Economic Prospects and the Developing Countries.* World Bank, Washington, D.C.

World Bank (1995b) *World Development Report 1995.* World Bank, Washington, D.C.

World Resources Institute (1994) *World Resources 1994–1995.* Oxford University Press, Oxford, U.K.

Wu, K. and Li, B. (1995) Energy development in China. *Energy Policy,* 23(2): 167–178.

Zang, Z.X. (1995) Energy conservation in China. *Energy Policy,* 23(2): 159–166.

Index